New Movements in the Study and Teaching of Biology

New Movements in the Study and Teaching of Biology

Edited by CYRIL SELMES

Temple Smith · London

First published in Great Britain 1974 by
Maurice Temple Smith Ltd
37 Great Russell Street, London WC1

ISBN 0 8511 7049 8

Printed in Great Britain by
Billing & Sons Limited, Guildford and London

Contents

VALUES AND IDEALS

Introduction

CYRIL SELMES

Biology is still a rapidly expanding subject, both in the knowledge of living organisms which is continually growing and in the techniques by which this knowledge is obtained. Even before the 1940s it was becoming true that a specialist in one branch of biology would have difficulty in understanding a specialist in another. This situation continues to become more complex: as biological knowledge increases, it has become impossible to add this new knowledge to existing degree courses—although both university and school syllabuses still show signs of this approach—and universities have not only formed Schools of Biology in order to provide more integrated approaches to biology teaching but also to allow more choice of options to the students. This increasing choice and specialisation, however, leads to even greater difficulties in comprehension between different kinds of specialists.

These changes have also been reflected in the new biology courses for secondary schools, both in the BSCS projects of the USA and in the Nuffield Biology projects of the UK, where up-to-date knowledge and techniques have been combined with more emphasis on the process of biological investigation and greater encouragement of individual learning through experimental work and projects. At the same time, however, there has been a reduction in the amount of factual material which has to be retained by the individual for examinations, which themselves have changed in both purpose and style.

If the changes associated with these projects were widespread, the direction in which biology teaching is heading—at least in secondary schools—might be clearer. Only a minority of pupils, however, take the special examinations which are designed to assess their work in these courses. It must therefore be assumed that the vast majority of pupils do not benefit fully from these changes even if their teachers make use of many of the projects' experimental techniques.

The biology teacher is faced with an increasing range of demands by these curriculum projects, ranging from the need to acquire understanding of newer knowledge and techniques to the application of biology to the problems of conservation, pollution and sex education. In addition, new methods of teaching and evaluation are being advocated, as well as the integration of biology not only with other traditional sciences (chemistry and physics) but also with geography, geology, sociology and psychology in the form of environmental science.

This book has a three-fold purpose: to present information about new areas of knowledge in biology, together with suggestions about the influence this knowledge should have on the teaching of biology at both secondary school and university level; to consider new approaches to the imparting of knowledge, skills and attitudes, particularly in secondary schools; and finally, to consider some of the values and ideals which underlie the teaching and learning of biology.

These purposes are reflected in the arrangement of the book into three sections. In Section 1, specialist biologists have been asked to explain recent developments in their branch of biology and relate these to teaching.

Each writer focuses attention on recent developments (or advances) which are considered to be important ones. Some authors summarise basic ideas in their subject area, like Michael Balls on 'Developmental Biology', David Tomley on 'Population Genetics and Evolution', and Marian Dawkins on 'Behaviour and Neurophysiology'; some concentrate on one or two specific topics, Michael Tribe and Peter Whittaker on 'Biochemistry' and Alexander Geddes on 'Biophysics'; whilst Richard Joske devotes more space to the implications of recent research when writing about 'The New Medicine and the Teaching of Biology', and Wilfrid Dowdeswell devotes more space to the teaching of his specialism, 'Ecology and Conservation'. These differences not only show the various interests of the authors but also the wealth of new knowledge and the conflicting demands this makes on the education of university students of biology. These chapters will introduce teachers of biology to areas which are developing rapidly and the references supplied by each author should enable the reader to study them further. In these chapters several themes appear. First there is the hypothetical and interpretive nature of biological investigation, which provides much of the stimulus and enjoyment for the investigators. Then there is the emphasis on a multidisciplinary approach to the solution of biological problems, bringing together specialists of different kinds and cutting across subject disciplines. Thirdly, there is the caution with which specialists make generalisations and yet their concern is for the dissemination of their knowledge and the development of an informed community.

The implications for teaching at all levels would seem to be the need for a multidisciplinary and problem-solving approach, the importance of providing courses of basic biological knowledge, possibly based on human biology, and the need also of providing a variety of optional courses in which students may develop their own interests but through work which involves both cooperation with others and the study of environmental or social problems.

Section 2 contains eight chapters which are concerned with some of the new approaches to the teaching of biology in secondary schools. Each author has been closely involved in the work described. Doris Falk urges the case for careful definition of behavioural or instructional objectives with particular reference to biology; an approach which appears to have received little attention in Britain, and yet one which would appear particularly useful in devising the kind of independent learning described by Donald Reid and Philip Booth. Their work shows the need for careful analysis of subject matter and raises the intertwined problems of motivation, the role of the teacher, and organisation; although they conclude that using a variety of teaching methods is always of great importance. In a similar way Brian Dudley argues that a variety of mathematical techniques may be used to augment the coherence of what is taught, and may also help to identify biological problems. Another group of techniques, which helps to provide greater understanding of biological problems, involves the use of radioisotopes, and the advantages and problems of these techniques are described by David Hornsey. The emphasis on biological

problems is continued by Dennis Fox in writing about the
biology components of the Secondary Science project; he
also shows how the project augments the suggestion of the
Newsom Report that pupils need to 'get the feel of being a
scientist' and that biological knowledge must be relevant to
the everyday needs of the children. One aspect of biology
which is usually considered relevant to secondary-school
children is sex education, a topic discussed by Jane Jenks
who argues that recent developments have replaced preoccupation with reproductive anatomy and physiology by concern about human relationships as well as social and moral
education. All these new approaches place less emphasis on
the accumulation of biological knowledge by pupils and more
emphasis on understanding through investigation; some of
the problems involved in learning biology through these new
approaches are outlined by Jack Dunham who also shows the
interaction between attitudes, motivation and learning, and
states the need for research into classroom learning. The final
chapter in this section concerns the assessment of the
learning of students. Robert Lister not only argues that the
teacher should attempt to define in behavioural terms the
intellectual qualities he wishes to assess but also that the
methods of assessment must be closely related to the
objectives of the teaching.

Section 3 considers the values and ideals of biology, the
reasons why new approaches are necessary. In 'The Need for
Change', Colin Stoneman summarises both the need for
changes in biology teaching and the characteristics of the new
science-teaching projects, as well as discussing the effects on
children and teachers and on examining. Like previous
writers he also sees the need for research and evaluation of
the new methods and materials, and asserts that biology
teaching (like biological science) must concern itself with the
real, external world. In 'From Flowers to Human Ecology',
Hugh Iltis displays the joy obtained from a lifetime of
addiction to botany, together with his growing concern about
the environment in which we live—thus again emphasising
that biology teachers (and teaching) should involve themselves in the real, environmental problems which require
biological understanding for their solution. And in the final
chapter, 'Agents of Change?', I argue that biology teachers

need to clarify their aims (or goals) as these are the beliefs and assumptions upon which their teaching should be based, whether this involves the teaching of biology as a separate subject or an interdisciplinary involvement.

Finally, all the writers hope this book will make a useful contribution to the ever-changing ideas on the role and presentation of the biological sciences.

It is the supreme art of the teacher to awaken joy
in creative expression and knowledge.
Albert Einstein

New developments

Biochemistry

MICHAEL TRIBE and PETER WHITTAKER

Introduction

... The material nature of the genes is still incompletely established, nor is their mode of action clear, although it seems probable that they influence enzymes concerned in protein synthesis ... the highly polymerized desoxyribose nucleic acid ... seems to be essential for the reproduction of the genes ... (and) the amount of ribose nucleic acid in the cytoplasm of a cell appears to be related to its capacity to effect protein synthesis ...

From *Plant Form and Function*
F.E. Fritsch and E.J. Salisbury

An examination of A-level Biology textbooks which were in common use in 1953, reveals remarkably few references to, and only cursory mention of, either ATP, DNA or protein synthesis. In contrast, virtually all O-level and A-level textbooks written since 1965 contain extensive references to these important biological molecules. This is in no way a criticism of the older texts, nor is it extolling the virtues of the new, but rather it is an indication of the remarkable developments which have taken place in biological science during the last two decades. Those of us who have been involved with the problems of biology have been fortunate enough to witness major breakthroughs in our understanding about how life processes work at the molecular level. The stimulus of these findings, however, has produced many problems.

Faced with the wealth of new information which is published daily on DNA, protein synthesis, energetics, enzymes, hormones, viruses, antibodies, etc., the professional biochemist is hard put to keep abreast of the literature. What chance therefore has the teacher, with only limited access to a comprehensive biochemical library, of keeping up with recent advances in all areas of biochemistry?

In many respects biochemistry is a difficult subject, since it demands an understanding of physical, chemical and biological concepts. Indeed, developments in physics and chemistry have provided at least part of the foundation on which modern biology is built, and have since sustained its rapid growth by providing the sophisticated technology and instrumentation which have been required. Consequently, there has been growing concern amongst biochemists, biologists and medical scientists to examine the content and the methods of teaching and learning biochemistry in tertiary education. An outcome of this concern over the education of biochemists has been the publication in 1972 of a quarterly journal of *Biochemical Education*, whose major aim is 'to facilitate the international exchange of ideas and information between biochemists who are actively engaged in the teaching of their subject at a University level to both science and medical students . . . '

It is against this background that we have tried to highlight growth areas in biochemistry and from these we have selected two areas for more detailed consideration, both of which seem to us relevant to curricula in both biology and chemistry in secondary schools. In the course of our survey we have endeavoured to point out the major problems challenging biochemists today, although it is impossible to be comprehensive. Therefore, at the end of the chapter we have provided a general reading list in the areas mentioned, as well as a source list of more advanced references for those anxious to gain a deeper insight into a particular field of research.

After reviewing the growth areas of biochemistry we give our personal reasons why and how we think biochemistry should be taught; when it should be taught; and what kind of science teaching in schools provides the most suitable background for those wishing to pursue the subject at the tertiary level.

Major areas of growth
Certain areas of biochemistry stand out in our minds as regions where either rapid progress is being made or the problems present such important challenges that major efforts to answer them can be expected in the near future.

The mechanism of DNA replication in the cell, for example, has captured the interest of biochemists and biologists at all levels since Watson and Crick originally defined the role of this molecule in heredity in 1953. Although the general concepts are known by virtually all biologists, little is known of the biochemical details of the replication process itself or how DNA synthesis is controlled during the cell cycle. Another area of comparative ignorance is the organisation of DNA in eukaryotic chromosomes.

The expression of the information encoded in DNA is another field of intense biochemical activity. The transcription of the nucleotide sequence of DNA into a messenger RNA copy, and the precise translation of the messenger RNA sequence into an amino-acid sequence of a protein, often appear as cut and dried processes in general text-books. In fact we know little of the actual process of messenger RNA synthesis and its control, particularly in eukaryotes. More is known of the mechanism of protein synthesis, although this is far from being completely established. Major efforts are being concentrated on the control of protein synthesis as this is almost certainly crucial in three other growth points in biochemistry. These are the mechanisms of cell-differentiation, hormone action and antibody production.

Study of the biochemistry of cell differentiation is still in its infancy. Much of the published work in this field deals with biochemical changes in some relatively simple model systems, such as bacterial sporulation, morphogenesis of *Acetabularia*, or aggregation of amoeboid cells and fruiting body formation in cellular slime moulds. The development of frogs from tadpoles has also been extensively studied, particularly the production of the enzyme collagenase which brings about tail resorption by degradation of its collagen, and the change from nitrogen excretion as ammonia (suitable only for aquatic animals) to excretion as urea, which involves 'turning on' synthesis of the urea cycle enzymes. No common conceptual basis for differentiation processes has yet emerged. What is clear is that these are extremely complex, and involve delicate interplay of nuclear, cytoplasmic and environmental factors.

Apart from a possible role in controlling differentiation, hormones play a vital part in controlling cellular metabolism.

This control can operate on at least two levels—enzyme synthesis and enzyme activity. The steroid hormones, for example, bring about increased RNA and protein synthesis in their target tissues. The primary event is the formation of a complex between the hormone and a receptor protein in the cytoplasm. The complex is transferred to the nucleus where it activates synthesis of specific messenger RNAs, coding for those proteins characteristic of the target tissue. Present studies aim to define the nature of the receptor proteins and the precise action of the hormone-receptor complex. The importance of these studies in the fields of fertility and contraception are clear. Some of the other hormones, (for example adrenalin, glucagon and ACTH) activate the enzyme, adenyl cyclase, in the cell membranes of the target tissue. This enzyme brings about the conversion of ATP to cyclic AMP which acts as an intracellular messenger. By a rather complex series of events, the cyclic AMP in liver and muscle cells activates glycogen phosphorylase (which catalyses glycogen breakdown) and inactivates glycogen synthetase (responsible for glycogen synthesis). The result of this is that adrenalin or glucagon can bring about mobilisation of liver or muscle glycogen reserves.

The production of antibodies by the plasma cells of the vertebrate lymphoid system in response to an antigen is another area where specific protein synthesis is elicited by a specific trigger. The increase in synthesis of a particular antibody results from the proliferation of a particular lymphocyte clone, whose primary job is synthesis of just that one antibody. Research in this area is directed towards establishing the origin of the genetic information for the vast range of antibody proteins which can be produced, and the mechanism of stimulation by the antigen of proliferation of the relevant lymphocyte clone. A fuller understanding of these processes would obviously add much to our knowledge in such areas as resistance to infection, graft-rejection and allergies.

One important feature of the action of both antigens and some hormones is that their primary reaction site is at the cell membrane. The general importance of membranes in the regulation of cellular processes is only just becoming clear. Considerable efforts are being devoted to establishing the structures of membranes and the mechanisms of membrane bound enzymes which are responsible for controlling the flow of metabolites and establishing and maintaining concentration gradients. Such work has important repercussions in the field of nerve and muscle physiology.

ATP synthesis in chloroplasts and mitochondria is carried out by membrane-organised, multi-enzyme complexes whose role has been established for about twenty-five years, but whose reaction mechanism is still not established.

Finally, in the last few years, great strides have been made in our understanding of enzyme action. The discovery of regulatory (allosteric) enzymes, whose substrate affinity is modulated by other metabolites, has shown how fine regulation of metabolic pathways can occur. High resolution X-ray diffraction studies on a number of enzymes have allowed the construction of three-dimensional molecular models of these. With these models it has been possible to show the probable substrate site, and which of the amino-acids in the enzyme interacts with the substrate, and also to make predictions concerning the actual mechanism of enzyme catalysis.

Specific growth areas
At this stage we wish to focus on the current state of two areas of biochemical research activity—DNA replication and ATP synthesis. We have chosen these for a number of reasons. They are essential processes in all organisms, they are studied at all levels of biochemistry and biology, and their study illustrates some of the widely different approaches to investigations of biochemical processes as well as some of the challenges and problems which may be encountered. We do not wish to give the impression that these are the only important areas of biochemistry, but with limited space we have chosen two which particularly interest us.

DNA replication
In 1953 Watson and Crick proposed that DNA consisted of two polynucleotide chains coiled around a common axis in the form of a double helix (fig. 1). The backbone of each chain comprises alternating deoxyribose and phosphate residues. Attached to each deoxyribose residue is one of the

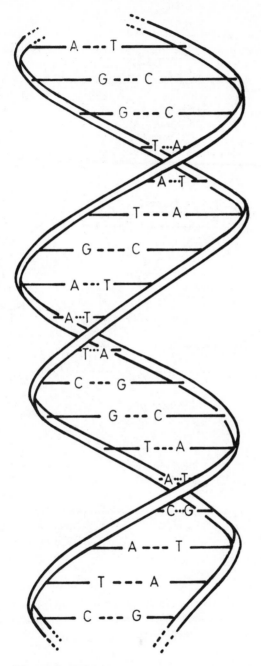

Figure 1 DNA structure

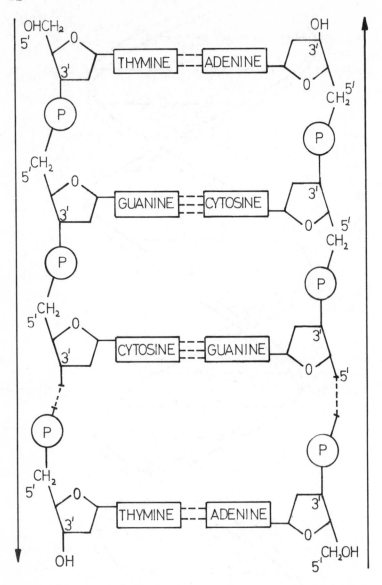

Figure 2 Portion of the DNA molecule showing antiparallel strands. Arrows indicate 5'→3' direction of the deoxyribose-phosphate linkages in each strand.

bases: adenine, guanine, cytosine and thymine. These point in towards the axis of the double helix and pair, by hydrogen bonding, with a complementary base on the other chain. The permitted base pairs are adenine-thymine (A–T) and guanine-cytosine (G–C). The two DNA chains run in an antiparallel direction. The phosphate residues join the hydroxyl on C atom 5′ of one deoxyribose with that on C atom 3′ of the adjacent deoxyribose residue (fig. 2). At one end (3′ end) of the chain there will be a deoxyribose with a free 3′ hydroxyl and at the other (5′ end) a deoxyribose with a free 5′ hydroxyl. One chain runs in the 3′ to 5′ direction, the accompanying strand in the 5′ to 3′ direction. Watson and Crick suggested that specific base pairing could allow for accurate replication of genetic information if each chain acted as a template for the synthesis of a complementary strand (fig. 3).

This concept was confirmed in 1958 by Meselson and Stahl. They grew the bacterium *Escherchia coli* on a medium which had NH_4Cl containing the ^{15}N isotope (rather than the normal ^{14}N isotope) as the sole source of nitrogen. DNA extracted from bacteria grown under these circumstances can be separated from normal DNA by centrifugation on a caesium chloride density gradient, because of its higher buoyant density. The cells were then transferred to a medium containing $^{14}N-NH_4Cl$ and at varying times of further

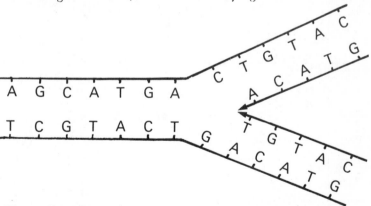

Figure 3 Watson and Crick's suggestion for the mechanism of DNA replication. Arrows indicate direction of synthesis of new DNA.

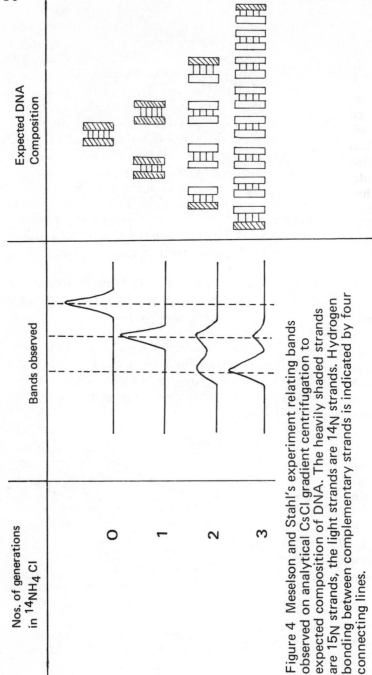

Figure 4 Meselson and Stahl's experiment relating bands observed on analytical CsCl gradient centrifugation to expected composition of DNA. The heavily shaded strands are 15N strands, the light strands are 14N strands. Hydrogen bonding between complementary strands is indicated by four connecting lines.

growth a sample was taken, the DNA extracted, and its composition determined by centrifugation on a caesium chloride density gradient (fig. 4).

After one generation the original ^{15}N–DNA had been completely replaced by a single species of density intermediate between that of ^{15}N–DNA and ^{14}N–DNA. After two generations half the DNA is ^{14}N–DNA and half is of the intermediate density. After three generations approximately three-quarters of the DNA is ^{14}N–DNA and one-quarter is of intermediate density. This result can be explained in terms of the Watson and Crick model for DNA replication, if the intermediate density DNA contains one strand of ^{15}N (old) DNA and one strand of ^{14}N (newly synthesised) DNA. Meselson and Stahl went on to demonstrate that this was in fact the case.

In 1963 further substance was given to the Watson and Crick model for DNA replication, when Cairns published autoradiographs of DNA extracted from *E. coli* grown on a medium containing radioactively labelled thymine. These clearly showed that a replication fork travelled right round the circular *E. coli* chromosome until two new circles were completed.

At the present time much work is being done to establish the enzymic mechanism of DNA replication at the molecular level. An important requirement of any biochemical investigation of this nature is a cell-free system capable of carrying out the activity under study, so that this can be studied without interference by other cellular processes. So far only limited progress has been made in this direction. Three different enzymes, capable of polymerising deoxyribonucleoside triphosphates to give DNA, have been discovered in *E. coli*. These have been called DNA polymerases I, II and III. With any of these the rate of polymerisation is much slower *in vitro* than that observed in the cell, so efforts are being made to find an improved cell-free system in which to study DNA replication.

The function of DNA polymerase II is at present unknown. Studies on a bacterial mutant unable to make DNA polymerase I have shown that this is not essential for DNA-replication and it is now thought to play a special role in either repair of damaged DNA or genetic recombination.

DNA polymerase III is thought to be involved in the DNA-replication process. All three of the polymerases have an important property in common; they cause polymerisation by addition of a nucleoside 5′ monophosphate (donated by a nucleoside triphosphate) to the 3′ end of a growing chain (fig. 5). This means that chain growth can occur only in

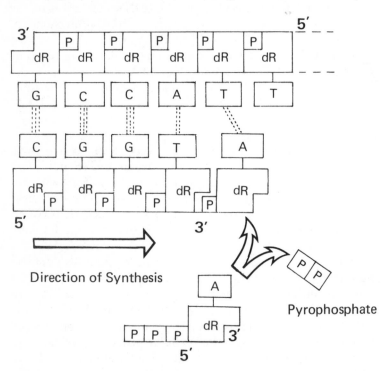

Figure 5 DNA synthesis in the 5′——▶3′ direction
dR = deoxyribose; P = phosphate; A,T,C,G, = Bases.

the 5′ to 3′ direction. This poses a technical problem for replication, as the method suggested by Watson and Crick implies replication of chains in two directions (fig. 6), one in the 5′ to 3′ direction and the other in the 3′ to 5′ direction. Up till the present time no enzyme capable of the latter type of replication has been discovered and the general view is that probably none exists.

Okazaki *et al.* (1968), have observed that newly

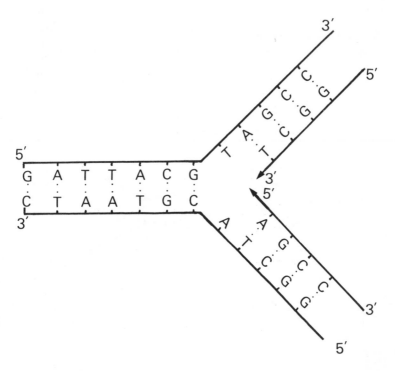

Figure 6 Initial suggestion for DNA replication showing DNA synthesis in 3'→5' and 5'→3' directions.

synthesised DNA is in very short strands and have suggested a method of DNA replication involving synthesis in only the 5' to 3' direction (fig. 7).

Replication of the molecule proceeds in the 5' to 3' direction along one strand (a and b fig. 7) for a short distance. The polymerase then moves to the other chain and replicates that in the 5' to 3' direction (b). A specific nuclease enzyme (the dotted arrow in b) then splits the newly synthesised chain. The polymerase then replicates another segment (c) on the two chains, which is similarly split. The newly synthesised fragments are subsequently linked together by the enzyme DNA ligase (d). Other similar theories have been suggested—these have in common synthesis of both chains by a single enzyme and discontinuities in DNA synthesis.

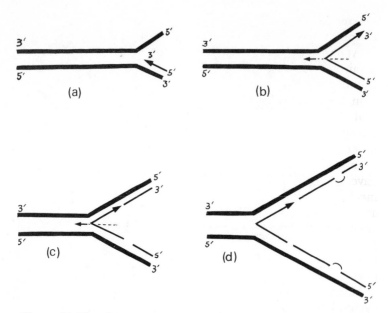

Figure 7 The Okazaki model for DNA replication. Thick
lines = old DNA. Thin lines = newly synthesised DNA.
Dotted arrow = site of nuclease action.

The synthesis of DNA presents many other technical
problems such as the mechanism of unwinding of the double
helix which is essential for replication. It can be calculated
that the rate of DNA replication involves a speed of
unwinding of more than 10,000 rpm. Furthermore, we are
almost completely ignorant of the molecular basis of DNA
replication in eukaryotes. This illustrates the enormous scope
which exists for further research in this important area.

Synthesis of ATP
The coupling of ATP synthesis (from ADP and inorganic
phosphate) to the oxidation of tricarboxylic acid cycle
intermediates (oxidative phosphorylation) was established
about 1940. The involvement in this process of the cytochrome system, rediscovered fifteen years earlier by Keilin,
was also suggested. In 1949 Kennedy and Lehninger discovered that oxidative phosphorylation occurred in mitochondria, although they knew little about the mechanism

whereby phosphorylation is coupled to oxidation.

At the present time we have a fairly detailed knowledge of the organisation of the electron transport chain; we know that electron transport and phosphorylation occur on the inner of the two membranes of mitochondria, but we still have not established the mechanism whereby oxidation and phosphorylation are coupled. Fig. 8 shows our present conception of the mitochondrial electron transport chain.

Malate, a tricarboxylic acid cycle intermediate, is oxidised to oxaloacetate, the oxidation being linked to the reduction of NAD^+ to NADH. NADH is in turn reoxidised to NAD^+ when linked to the reduction of flavoprotein. Similar oxidation-reduction reactions proceed along the elctron transport chain until O_2 is reduced.

This stepwise oxidation of malate (or other metabolite) allows the controlled release of the oxidation energy. At certain of the oxidation-reduction steps sufficient energy is available for the synthesis of an ATP molecule, as shown in fig. 8. The coupling of electron transport and phosphorylation is not as casual as this, however.

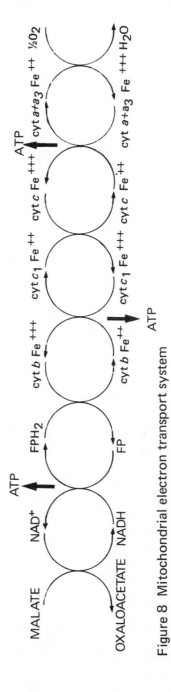

Figure 8 Mitochondrial electron transport system

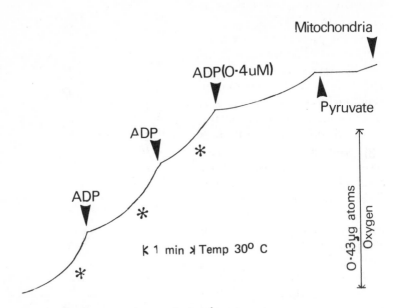

Figure 9 Oxygen electrode tracing

Fig. 9 shows that on addition of a substrate, such as pyruvate, to insect flight-muscle mitochondria, only a slow rate of oxygen uptake is observed. If a small quantity of ADP is added, rapid O_2 uptake occurs until all the ADP has been converted to ATP (position marked * on fig. 9). A further addition of ADP gives another burst of oxygen uptake. This shows that oxidation is negligible unless phosphorylation is taking place at the same time. In other words, oxidation and phosphorylation are obligatorily coupled.

It was natural for biochemists to think in the first place of a chemical reaction sequence for the coupling process. One such suggestion for ATP synthesis at the cyt $b \rightarrow$ cyt c_1 stage of electron transport is shown in fig. 10.

The oxidation of cytochrome b by cytochrome c_1 is considered to provide the energy for the synthesis of a complex (sometimes referred to as a 'high-energy' intermediate) cyt $b - I$ (1). Cyt $b - I$ then donates I to an enzyme E which is involved in the terminal steps of phosphorylation

(1) cyt b Fe^{++} + cyt c_1 Fe^{+++} + I$\rightleftharpoons$ cyt b Fe^{+++} ——I + cyt c_1 Fe^{++}

(2) cyt b Fe^{+++} —— I + E $\rightleftharpoons$ cyt b Fe^{+++} + E–I

(3) E–I + phosphate $\rightleftharpoons$ E–phosphate + I

(4) E —— phosphate + ADP $\rightleftharpoons$ E + ATP

Figure 10 A possible chemical coupling mechanism for oxidative phosphorylation

(2). A phosphorylated enzyme is then formed (3) which can donate phosphate to ADP (4). This is not the only chemical sequence that can be written for the coupling process, nor is it the only one which has been written. All have in common the formation by electron carriers of intermediate complexes such as cyt $b - $I. Although intensive efforts have been made to detect and isolate such intermediates, no convincing demonstration has yet emerged. This failure, and the necessity for an intact membrane structure for ATP synthesis, led Mitchell (1961) to propose an alternative scheme for ATP synthesis: namely the chemiosmotic coupling hypothesis. Mitchell suggests that operation of the electron transport system results in the establishment of a pH differential and membrane potential, across the inner membrane of the mitochondrion (fig. 11).

At certain points in the electron transport system, reaction occurs between a hydrogen carrier (e.g. FP) and an electron carrier (e.g. any cytochrome). Coupled oxidation of reduced flavoprotein by oxidised cytochrome b, for example, results in the release of protons. These are considered to be released in a particular direction (outwards). Oxidation of reduced cytochrome $(a + a_3)$ by molecular oxygen requires protons, and

22

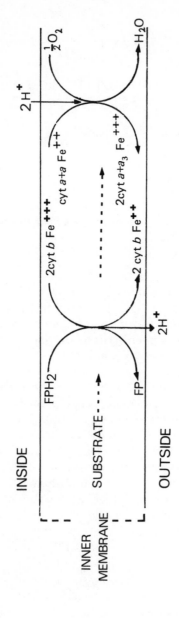

Figure 11 Chemiosmotic hypothesis. Establishment of a proton gradient by the electron transport system

these are considered to come into the membrane from inside the mitochondrion. These effects are achieved by virtue of a precise spatial orientation of the electron transport system in the membrane. If, as the theory requires, the membrane is impermeable to H^+, operation of the electron transport system would lead to the establishment of a hydrogen ion gradient and a membrane potential. Mitchell considers that this is how the energy made available by electron transport is trapped initially.

The hydrogen ion gradient and membrane potential can then, according to Mitchell, be used to drive ATP synthesis from ADP and phosphate. He considers that this is carried out by an enzyme, adenosine triphosphatase (ATPase), which is located in the membrane (fig. 12). Under normal, aqueous conditions an ATPase would be expected to bring about the hydrolysis of ATP to ADP and phosphate as follows:

$$ATP + H_2O \longrightarrow ADP + H_3PO_4$$

This enzyme is considered to be located in a very

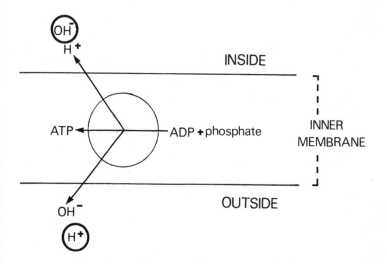

Figure 12 Chemiosmotic hypothesis. Proton gradient 'pulls' ATPase in direction of ATP synthesis

hydrophobic environment within the inner membrane. In other words, the concentration of water at the enzyme's active site is going to be very low and so the rate of hydrolysis of ATP must be low. The excess of H^+ which was pumped outside by the electron transport system draws OH^- from the membrane. Similarly the deficit of H^+ (= excess of OH^-) inside draws H^+ out of the membrane. This effectively lowers the concentration of water at the active site of the enzyme and causes ATP synthesis by a reversal of ATPase action.

Recently a third coupling hypothesis—the conformational coupling hypothesis—has been put forward. This proposes that oxidation energy is trapped as a conformational change of the mitochondrial inner membrane, and that the reverse conformational change drives ATP synthesis. The theory is based on the conformational differences observed from electron micrographs of the inner mitochondrial membrane, when little or no ADP is present (resting or orthodox state) and on the addition of excess ADP (active or contracted state), see fig. 13.

No compelling evidence has emerged in favour of any of these coupling theories. Evidence proposed by the proponents of one theory can invariably be explained away by devotees of the others. It is not easy to predict the direction of progress in this field as most of the obvious avenues of investigation have been explored. Recently a number of groups have begun to investigate the properties of oxidative phosphorylation-deficient mutants of yeast. It is possible that an oblique attack such as this could prove fruitful.

Developments in the teaching of biochemistry at University level

We have outlined where some of the major areas of advance in biochemistry lie. We now need to examine a number of important questions concerning the effect that this rapidly growing subject will have on biochemistry teaching in particular and science teaching in general.

It seems appropriate therefore to centre our discussion around three questions:

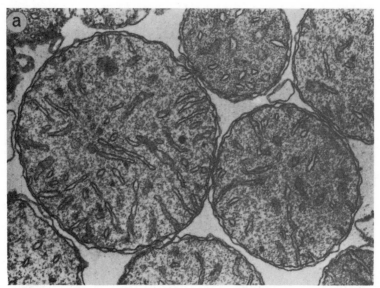

Figure 13
(a) Orthodox configuration: ADP uniting: minimal electron flow
(b) Contracted configuration: ADP in excess: maximal electron flow

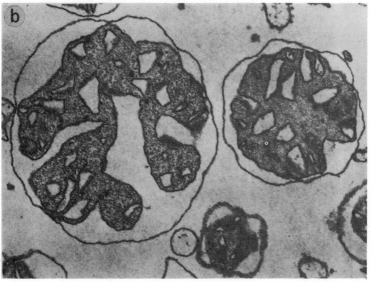

Scale   0·5μm

1 What are the basic aims of biochemistry teaching?
2 To whom are we teaching biochemistry and what are their specific requirements?
3 What range of teaching and learning experiences should we be providing in biochemistry at the tertiary level?

With reference to question 1, an examination of undergraduate biochemistry courses shows them to be organised into six major areas, although different universities, as is their prerogative, emphasise some areas more than others. These areas are: biological macromolecules, enzymes, molecular genetics and biosynthesis, cell energetics, metabolic pathways, and regulation of cell processes.

Within these broad areas we feel that the teaching of biochemistry should aim:

1 To convey important biochemical concepts and principles in an analytical and contemporary context, so that information and ideas are supported as far as possible by experimental evidence.
2 To develop the intellectual and practical abilities of students which are fundamental to an understanding of science in general and biochemistry in particular.
3 To develop in students the facility to work and think independently, and particularly how to learn through reading and critical evaluation of ideas.
4 To develop in students an understanding of the processes involved in biochemical research and the application and implications of this research for society.
5 To develop in students an ability to communicate effectively the concepts and importance of biochemical research to others.

Amongst the intellectual and practical abilities suggested in aim 2 above we would include:

(i)   a basic knowledge of biochemical facts and vocabulary;
(ii) an understanding of basic biological and chemical concepts;
(iii) an ability to formulate hypotheses or deduce meanings from a given set of observations;
(iv) an ability to devise experiments to test hypotheses

and evaluate data in support of hypotheses;
(v) an ability to trace or follow up problems through the systematic use of scientific literature;
(vi) learning efficient study habits;
(vii) making value judgements.

Although some readers will not agree with all the aims put forward in this chapter, and we certainly cannot claim to have achieved all these aims with our own students, the aims do have many points in common with recent innovations in curriculum development (e.g. Nuffield Science and the Open University Science Courses), so that we feel that they reflect contemporary opinion. In this respect too, we welcome the interesting approach adopted by Kerridge and Tipton (1972) in their text book for undergraduates, *Biochemical Reasoning*. The book has a number of eminent contributors with various research interests, and deals with a collection of data-handling problems ranging from fairly elementary to quite advanced standards.

Regarding the second question: there are three major groups of students who will be involved in biochemistry courses at the tertiary level:

(i) those who intend to go on to biochemical research and/or teaching;
(ii) those who require biochemistry as a 'service' course for other branches of science, e.g. medicine, food technology, etc;
(iii) those students who have a general interest in bio- chemistry, but who do not intend to use the subject directly in a vocational sense.

At present there are comparatively few students who fall into this last category, but biochemists should face up to the possibility that the numbers of such students may well increase over the next few years, and that existing courses may not be appropriate. Where the three categories of students co-exist in a university, it is customary to find that they are taught together in the first year (certainly in so far as lectures and practicals are concerned), despite the fact that their immediate and potential interests are different. Many biochemists may think this perfectly reasonable, whereas others, including ourselves, would argue that more 'tailor-made' courses are needed, i.e. that universities should increase the range of teaching and learning experiences to cater more for individual requirements in *all* courses, and by individual requirements we mean those both of teacher and taught. However, we realise that it is difficult to reconcile ideals with practicalities; there are logistic, resource, space and organisational problems to consider. Nevertheless, if universities are convinced that courses should be providing more for the individual, it will be necessary to modify the teaching pattern.

This then brings us on to question 3—what range of teaching and learning experiences should we be providing in biochemistry at the tertiary level?

For many years now, the lecture has been the principal teaching method in science courses. Lectures usually define the course and determine the examinations which follow. The result is that students feel compelled to attend lectures, not always for their instructional value, but often because they are afraid to miss something in which they will be examined. In his book, *What's the use of lectures?*, Donald Bligh (1971) finds that although the lecture is effective as a method for transmitting information and can at times awaken critical skills in students, it is not effective as an active method for promoting thought or changing attitudes. He therefore poses the question 'How can other methods be combined with lecturing?' if the teacher has a variety of objectives which cannot be achieved by lecturing. Certainly we are not advocating the abolition of lectures, but all of us involved in science education should seriously consider reducing the quantity, if need be, and increasing the quality of lectures. By doing so there would be more time available for the 'other methods'. It will not be possible simply to add other methods of teaching and learning to an already over-crowded timetable of lectures and practicals. The principal teaching method at the universities of Oxford and Cambridge is the tutorial system. Many universities, including our own, have adopted this system, particularly in arts subjects. The tutorial is certainly an extremely good form of teaching and learning, if effectively used and fully exploited. Far too often, however, it is used in science merely as an adjunct to the lecture course. Whilst admitting that this could be one of its roles, we would advocate its more extensive use for a variety of small group learning situations. For example, we can recommend the integration of tutorials with practical work to introduce, monitor and conclude a short, or more extensive, piece of project work. During the first tutorial the problem or hypothesis is presented and discussion ensues about the ways it might be solved experimentally. The experimental work then follows over one or more days, depending on the problem and available time, with one or more monitoring tutorials interposed to analyse the progress being made by the group. At the end of the experimental period the tutorial is used to analyse results and suggest where further experimentation is required. Later tutorials can be used to open up discussion on the wider issues of the problem, perhaps in relation to one or two selected papers. Students can then go away with a reading list to produce an essay on their experimental results in the light of published data. The tutorial therefore provides an opportunity of showing where biochemistry can help to explain biological problems at the level of the cell, whole organism or population.

Another approach is that advocated by Epstein (1970), in which some six tutorials are used to answer questions, such as 'How does the scientist think and work?' and 'How are important discoveries made?' At each tutorial, one of a number of papers, in chronological order, is presented to the students by the tutor; the papers having been previously selected to tell a scientific story in an important area of biochemistry. The introduction of subsequent papers is in turn determined by the 'experimental' outcome of the previous paper. One example, cited and used by Epstein, shows how DNA was established as the basic hereditary material; the papers selected include those of Delbrück (1940); Avery, MacLeod and McCarty (1944); Hershey and Chase (1952); Chargaff (1950); Watson and Crick (1953a) Wilkins, Stokes and Wilson (1953) and Meselson and Stahl (1958).

Since 1969, Sussex University along with four other universities, has been involved in the Inter-University Biology Teaching Project. One of the major aims of this project has

been the production of self-instructional material in programmed form (see Tribe, 1972). In deciding the course
content, we concentrated initially on developing a conceptual
structure on which the students can build more advanced
courses and to which more detailed information can be
attached later. The course structure has a significant biochemical component and can be used either as a basic course,
or to prime or reinforce existing lecture courses, especially
for students requiring a 'service' course, or those with a poor
biological background. In the future, we see more extensive
use being made of self-instructional material, as student
numbers increase and courses become more diverse.

Another problem of concern is the comparative lack of
opportunity for students to participate in group learning and
decision-making experiences. To this end we have designed a
number of simulation exercises, Tribe and Peacock (1973).
These exercises have been enthusiastically received by
students and have generated interest from Faculty. We look
forward to the introduction of other innovations in the
future; in particular, greater consultation between student
and tutor in the planning of each student's personal
work-schedule.

How can the schools help in the education of biochemists?
As the areas of biochemical knowledge expand, the demands
made by the universities and colleges on the schools for
students with a sound, broadly-based scientific education
increases. We do *not*, however, advocate that biochemistry
should be taught as a separate A-level subject at school. If
this were to happen, biochemistry might become too limited
and could easily lose sight of the wider biological problems.
Instead, we would recommend a combination of subjects
chosen from physical science, chemistry, physics, mathematics, physics with mathematics and biology. It is within
the areas of biology and chemistry that we would think it
appropriate to introduce aspects of biochemistry. For
example, in the Nuffield A-level chemistry course, all
students are required to follow nineteen topics together with
one special study to be selected from five alternatives; one of
these alternatives is biochemistry. As is pointed out in the
introduction to the special study, two of the aims are to

appreciate 'their relevance to neighbouring subjects, and for learning something of the social and economic effects of chemistry.'

There are, indeed, two very good chapters in the special study, which examine 'Biochemistry and Commerce' and 'Agricultural aspects of Biochemistry'. We attach great importance to applied biochemistry and welcome the prominence given to this in Nuffield science courses and in recent articles in the *School Science Review* . It is important that all pupils (even those not going on to higher education) receive an education which includes some knowledge of environmental biochemistry; for example, aspects of sewage decomposition, recycling of organic materials and the problems of non-biodegradable substances; the effects of herbicides and insecticides; fermentation; birth control and the principles of immunity.

In contrast to the special study in biochemistry, Nuffield biological science has preferred to disperse the biochemical aspects throughout the course, thus providing a broad perspective to biological problems by an alternative approach.

As stated earlier, biochemistry would not be in its present position without chemistry and physics. It is important therefore to maintain and improve communication, understanding and cooperation between biologists, chemists and physicists, for there is a great danger of all scientists becoming more insular as scientific knowledge expands.

Let us close this chapter by citing one recent example, which has caused considerable controversy amongst biologists and chemists, and where greater mutual understanding is necessary: namely, the idea, originally proposed by Lipmann in 1941, that ATP is a 'high energy' compound.

It is unfortunate that the formula of ATP is frequently abbreviated to Adenine-Ribose-P~P~P, because the 'squiggle' notations (~) used to distinguish between the interphosphate linkages and the ribose-phosphate linkage of ATP, earlier came to be regarded as 'high energy bonds'. This is a total misconception. The idea that energy can be stored in the squiggle bond (a covalent bond) is obviously nonsense because energy is always required to break covalent bonds. Some authors replace the term high energy bond with the term 'high energy molecule', meaning that ATP releases more energy on hydrolysis to ADP and phosphate ($\triangle G^\circ$ = −30 kJ/mole) than can be obtained, for example, on hydrolysis of glucose−6−phosphate to glucose and phosphate ($\triangle G^\circ$ = −13 kJ/mole). The implication is that the 'high energy' is not contained in a single bond, but is a function of the molecule as a whole, in relation to the other reactants and products participating in the reaction. Strictly speaking, therefore, in the reaction ATP + HOH $\longrightarrow$ ADP +$H_3 PO_4$, water can also be regarded as a high energy molecule. Again, the term high energy molecule, as applied to ATP and other organic phosphate esters, is only relative. In a series of organic substances which take part in phosphate group transfer, some molecules, like phosphoenol-pyruvate, liberate more free energy on hydrolysis than ATP (see table 1).

Table 1 Standard free energy of hydrolysis of phosphate compounds

	$\triangle G^\circ$ kJ/mole
Phosphoenol-pyruvate	− 62
Phosphocreatine	− 43
ATP (to ADP and phosphate)	− 30
Glucose − 1 − phosphate	− 21
Glucose − 6 − phosphate	− 14

($\triangle G^\circ$ is the standard free energy change of a reaction i.e. pH = 7.0; Temp. = 25°C; 1 M concentrations of reactants and products).

The term high energy molecule can also be criticised (Banks, 1970). The problem really arises from the difficulty that biochemists have encountered in finding a concise terminology to explain the important role of ATP in the cell. ATP does play a special role as an energy 'currency' (cf. Banks, 1970), as its effective hydrolysis to ADP and phosphate (although not in a single reaction) does provide the energy for muscular contraction, active transport across membranes, cell division etc. Also, a special organelle, the mitochondrion, has evolved, whose primary role is ATP synthesis from ADP and phosphate. In none of these processes can the role of ATP be exactly duplicated by another molecule. However, it is important that attempts to display biochemical concepts in a 'pictorial' fashion do not propagate inaccuracies.

There are two further criticisms which are perfectly valid and should be mentioned. The first concerns the use of $\Delta G°$ for ATP in a cellular context, where pH, temperature and concentration of reactants and products are not standard. The actual free energy change (ΔG) *in vivo* may therefore be very different. The second criticism is the use of the free energy concept, which is appropriate for a closed system (one in which energy is exchanged with the surroundings, but not matter) but inappropriate for living organisms which can be regarded as open systems (those in which both energy and matter are exchanged with the surroundings) in a steady state.

In future therefore, biochemists may have to reconsider how they formulate ATP's role in the cell. It may be better to play down the thermodynamic aspects and concentrate on the extremely important role fulfilled by ATP in its participation in linked reactions where phosphate group transfer takes place. ATP might be better regarded as a 'co-enzyme', selected in the course of evolution for its unique role in a great variety of enzymic reactions.

References

O. T. Avery, C. M. MacLeod, and M. McCarty (1944) 'Studies on the chemical nature of the substance inducing transformation of pneumococcal types I. Induction of transformation by a deoxyribonucleic acid fraction isolated from pneumococcus type III.', *Journal of Experimental Medicine*, 79.

B. E. C. Banks (1970) 'A misapplication of chemistry in biology', *School Science Review*, vol. 179.

D. A. Bligh (1971) *What's the use of lectures?*, University Teaching Methods Unit,London.

J. Cairns (1963) 'The bacterial chromosome and its manner of replicating as seen by autoradiography,' *Journal of Molecular Biology*, 6.

M. Delbrück (1940)'The growth of bacteriophage and lysis of the host', *Journal of General Physiology*, 23.

H. T. Epstein (1970) *A strategy for education*, Oxford University Press.

A. D. Hershey and M. Chase (1952) 'Independent functions of viral protein and nucleic acid in growth of bacteriophage', *Journal of General Physiology*, 36.

D. Keilin (1925) 'On cytochrome, a respiratory pigment, common to animals, yeast, and higher plants', *Proceedings of the Royal Society, Series B*.

E. P. Kennedy and A. L. Lehninger (1949) 'Oxidation of fatty acids and tricarboxylic acid cycle intermediates by isolated rat liver mitochondria', *Journal of Biological Chemistry*, 179.

D. Kerridge and K. Tipton (1972) *Biochemical Reasoning*, W. A. Benjamin Inc.

F. Lipmann (1941) 'Metabolic generation and utilisation of phosphate bond energy,' *Advanced Enzymology*, 1.

M. Meselson and F. W. Stahl (1958) 'The replication of DNA in *Escherichia coli*', *Proceedings of the National Academy of Science*, 44.

P. Mitchell (1961) 'Coupling of phosphorylation to electron and hydrogen transfer by a chemiosmotic type of mechanism', *Nature*, 191.

R. T. Okazaki, K. Okazaki, K. Sakabe, K. Sugimoto and A. Sugino (1968) 'Mechanism of DNA chain growth. I. Possible discontinuity and unusual secondary structure of newly synthesized chains', *Proceedings of the National Academy of Science*, 59.

M. A. Tribe (1972) 'Designing an introductory programmed course in Biology for undergraduates,' in *Aspects of Educational Technology VI* (Eds. Austwick, K. and Harris, N.D.C.).

M. A. Tribe and D. Peacock (1973) 'The use of simulated exercises (games) in biological education at the tertiary level, to be published in *Aspects of Educational Technology VII*.

J. D. Watson and F. H. C. Crick (1953) 'A structure for deoxyribonucleic acid', *Nature*, 171.

J. D. Watson and F. H. C. Crick (1953) 'Genetical implications of the structure of deoxyribonucleic acid', *Nature*, 171.

M. F. H. Wilkins, A. R. Stokes and H. R. Wilson (1953) 'Molecular structure of deoxypentose nucleic acids', *Nature*, 171.

Further Reading

ATP Synthesis–Oxidative Phosphorylation
R. P. E. Gregory (1971) *Biochemistry of Photosynthesis*, John Wiley & Son.
D. O. Hall and K. K. Rao (1972) *Photosynthesis. Studies in Biology* 37, Edward Arnold.
A. L. Lehninger (1971) *Bioenergetics*, W. A. Benjamin Inc.
A. L. Lehninger (1970) *Biochemistry*, Worth Publishers Inc.
M. A. Tribe and P. A. Whittaker (1972) *Chloroplasts and Mitochondria. Studies in Biology* 31, Edward Arnold.

DNA replication, RNA and protein synthesis
J. J. Davidson (1972) *The Biochemistry of the Nucleic Acids* (7th edition), Chapman & Hall and Science Paperbacks.
G. Stent (1971) *Molecular Genetics*, W. H. Freeman & Co.
J. Watson (1970) *Molecular Biology of the Gene* (2nd edition), W. A. Benjamin Inc.

Differentiation
M. Hamburgh (1971) *Theories of development*, Edward Arnold.
C. A. Pasternak (1970) *Biochemistry of Differentiation*, John Wiley & Sons.
M. Sussman (1964) *Growth & Development* (2nd edition), Prentice-Hall Inc.

Enzymes
S. Bernhard (1969) *Enzymes: Structure & Function*, W. A. Benjamin Inc.
R. E. Dickerson and I. Geis (1969) *The structure and action of Proteins*, Harper & Row.
D. C. Phillips (1966) 'The three-dimensional Structure of an Enzyme', *Scientific American*, 215.
A. Wiseman and B. J. Gould (1971) *Enzymes, their nature and role*, Hutchinson Educational.
F. Wold (1971) *Macromolecules; structure and function*, Prentice-Hall.

Hormones
A. G. Clegg and P. C. Clegg (1969) *Hormones, cells and organisms*, Heinemann Educational Ltd.
F. J. Ebling and K. C. Highnam (1969) *Chemical Communication, Studies in Biology*, 19, Edward Arnold.

Immunology
G. J. V. Nossal (1971) *Antibodies and Immunity*, Penguin Books.

Membranes
D. Branton (1969) 'Membrane Structure', *Annual Review of Plant Physiology*, 20.
A. P. M. Lockwood (1971) 'The Membranes of animal cells', *Studies in*

Biology 27, Edward Arnold.
D. H. Northcote (1968) 'Structure and function of membranes', *British Medical Bulletin*, 24.

Advanced sources of reference
Advances in Enzymology
Advances in Enzyme Regulation
Advances in Immunology
Advances in Protein Chemistry
Annual Review of Biochemistry
Annual Review of Microbiology
Biochim. Biophys. Acta Reviews
Biological Reviews
Cold Spring Harbour Symposia in Quantitative Biology
Essays in Biochemistry
Nature, New Biology
Physiological Reviews
Progress In Nucleic Acid Research & Molecular Biology
Society for Experimental Biology Symposia

Biophysics

ALEXANDER GEDDES

What is Biophysics?

There are probably as many definitions of biophysics as there are biophysicists and the reason for this is not too hard to discover. Biophysics is not so much a field of study as an approach, a way of thinking about problems in biology. It is concerned essentially with interpreting the properties of living systems in terms of the fundamental molecular events which are taking place. It differs from biochemistry in that it has a special interest in the *spatial* distribution of atoms within molecules, of molecules within macromolecular assemblies, and of these assemblies within the functioning unit.

Its excitement lies in its aim towards as complete an understanding of life processes as the intellect and technology will allow. To achieve this aim it brings together the biologist, physicist, chemist, and mathematician; for only by this unification of disciplines can the various levels of atomic, molecular, and cellular organisation be linked.

Recent developments

In describing recent progress it is difficult to know where to define the boundaries. Of undoubted interest are such varied topics as holographic models of memory storage and molecular orbital studies on hormones, but at the present time many of these fringe studies would not seem to have an immediate relevance to school biology. The subjects chosen for discussion below have been selected on the basis of the author's own interest and because they constitute a theme. Each is concerned with demonstrating the relevance of three-dimensional molecular structure to some wider aspect of biology.

Perhaps a word or two should be said first about instrumentation. Sometimes it seems that biophysicists lose themselves in a mist of technology and, although this does happen occasionally, they are still biologists at heart. Their machines are just tools, like the optical microscope is to the zoologist and the chromatography column to the biochemist. However, it cannot be denied that part of the biophysicist's role is to develop instrumentation to such an extent and in such a way that it can be usefully applied in the other more traditional branches of biology. In this respect we can mention the freeze-fracturing and scanning electron microscopes which are already finding wide application in such fields as cancer research and food science. There are a number of spectroscopic techniques where improved resolution and methodology has brought them within the range of biological problems. These include, for example, proton magnetic resonance and electron spin resonance where radio and microwave radiation respectively interact with the specimen in a high magnetic field so as to produce a spectrum in which the absorption frequencies depend on the chemical environments of protons or unpaired electrons. In the latter case the unpaired electrons might occur in transition metals present in metallo-enzymes or in stable free radicals attached as labels to some important metabolite. With their enhanced intensity and subsequent ease of measurement laser beams are finding application in light-scattering experiments used to determine particle size and shape, or to provide information about the vibrations and energies of chemical bonds. Because biological problems are so complex, intensive use is made of computers both to analyse the data and to control the machinery. This is particularly so in X-ray crystallography which used to be such a time-consuming and often soul-destroying technique. Use is also being made of nuclear reactors and particle accelerators, and with such power available we can even study the molecular fibres of a muscle while it is in the process of contraction.

Evolution: molecules and elephants' trunks
Whereas the gross morphological features of an organism lie prey to the rigours of natural selection, the changing structure of a molecule can more closely reflect the underlying rate of genetic mutation. In particular, a study of the species variation in the amino-acid sequence of proteins having identical functions has provided a fascinating insight

into the mechanisms of evolution.

One of the first things we notice is that each protein changes (evolves) at a constant rate, and it is possible to define a 'unit evolutionary period' which is the time taken for a 1 per cent change in sequence between two diverging lines. This period varies from protein to protein, so that we are not measuring the true DNA mutation rate but rather the rate of 'harmless' mutations, and this rate will depend on how seriously the mutation affects the function of the protein. Another quite startling fact to emerge is that some proteins can change up to 50 per cent or so of their amino-acids and yet still retain their ability to perform highly specialised functions. Some parts of the sequence are always invariant and these amino-acids are obviously vital to the protein's activity. Some amino-acids can be changed 'conservatively' i.e. replaced by similar residues, while a few can sometimes be replaced randomly.

Perhaps the most extensively studied protein, from the point of view of species sequence variation, is the enzyme cytochrome *c*. This enzyme forms part of the electron transport chain in mitochondria whereby electrons removed from foodstuffs are passed from molecule to molecule in a series of energy releasing steps, finally to meet with oxygen which can then combine with hydrogen ions to form water. It has an evolutionary period of about 20 million years and, of the 104 amino-acid residues in human cytochrome *c*, gradual changes can be traced back through the evolutionary tree until about 40 per cent of the molecule is different— some 1,000 million years, or about the time when plants and animals diverged.

A recent X-ray crystallographic determination of the three-dimensional molecular structure of the oxidised form of cytochrome *c* shows in splendid detail the reasons why some sequence sites are so much more important than others (Dickerson *et al.*, 1971). It is a roughly spherical molecule with a planar haem group containing its central iron atom (presumably essential for the electron transport process) sitting in a crevice with one edge exposed. It follows the general pattern of all protein molecules whose three-dimensional structures have been determined that the polar residues occur on the surface, where they can interact with an aqueous environment, and the non-polar residues are present in the interior, where they can stabilise the folding of the peptide backbone of the protein by so-called hydrophobic interactions. This arrangement accounts generally for the conservatism of most replacements. One exception to this rule is *Phenylalanine 82*, an invariant non-polar residue sitting on the surface near the haem crevice. The importance of this site became apparent when the structure of the reduced form of the molecule was obtained (Takano *et al.*, 1972), and here the phenylalanine ring has moved right into the crevice, cutting off or modifying electron access to the haem group. This movement implies that this residue is involved in the oxidation-reduction process and the necessity for its presence here proves more important than the energy destabilisation caused by having a non-polar group at the surface.

Many of the glycine residues are invariant and the reason for this is that their lack of a bulky side chain allows the peptide backbone to fold back sharply on itself without steric hindrance. This increase in folding possibilities allows what is a relatively small protein to get a really good hold on the haem group. The residues which actually bind the haem group are also invariant. Most of the other invariant residues seem to be contained in two spatially well-defined regions and it would seem reasonable to suppose that these regions are those parts of the molecule that interact on the one hand with the cytochrome b-c_1 reductase complex, which donates electrons to cytochrome c, and on the other hand with the cytochrome a-a_3 complex, which accepts electrons from cytochrome c. Certainly the electrostatic nature of one of these regions is compatible with other evidence concerning the binding of the oxidase complex.

We can see through these crystallographic studies that the essential properties of this important molecule are relatively unchanged through millions of years of evolution. The harmless mutations are those which do not seriously affect that particular folding pattern of the molecule which presents certain invariant residues at specific locations in three-dimensional space.

Apart from sequence changes there are other aspects of protein structure which might have an evolutionary basis. In the last few years it has been noticed that the larger proteins would seem to be organised in several spatially distinct regions, and that the 'active site' lies at the junction of these component parts. It may be that these parts are the remnants of simpler molecules which have aggregated through an evolutionary process into the larger more complex molecule.

Nucleic acids: What comes after the Double Helix?

The determination of the double helical nature of DNA, with all its implications for replication and transcription, will remain one of the most important achievements of the biophysical approach. With the recent publication of the complete three-dimensional structure of yeast phenylalanine Transfer RNA (Kim *et al.*, 1973), yet another hurdle in this field of gene transcription and protein assembly has now been overcome.

Each amino-acid has a corresponding transfer RNA (tRNA) molecule which transports it from the cytoplasm to the ribosome/messenger RNA (mRNA) complex. The tRNA recognises and attaches itself to a particular triplet of bases (codon) in the mRNA, whose complete base sequence has been specified by the parent DNA. Starting at one end of the mRNA the various tRNA molecules arrive, deposit their amino-acid, and move on to collect another amino-acid. As each amino-acid arrives it is joined to the previous one until the protein is complete. It is obviously of great interest to know the shape of the tRNA molecule, especially the relationship between the amino-acid binding site and the codon recognition site or 'anticodon'. A great deal of effort has been put into molecular model building studies since these were particularly successful in helping to solve the structure of DNA itself. Nucleotide sequence studies of tRNA suggest that the molecular chain can aggregate in four places to form short lengths of base-paired double helix and most of the proposed models have incorporated these regions into a 'clover leaf'-shaped molecule. Sadly it turns out that this is not the correct structure—just to remind us that Nature will not so easily reveal her secrets. After considerable difficulties in the preparation of suitable crystals, X-ray diffraction studies now reveal an L-shaped molecule (fig. 1). It is rather like a hairpin whose prongs have been twisted around each other and the whole then bent in the middle

Diagrammatic representation of a tRNA molecule

1 The linear tRNA polymer (A=Adenine, C=Cytosine, G=Guanine, U=Uracil)
2 Folded into a hairpin shape, Regions of bases which sequence studies
   indicate might be involved in base pairing are shaded in like manner.
3 Folded into a clover leaf shape so that base pairing regions are aligned.

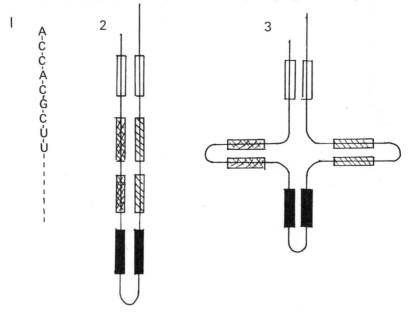

4 Formation of the base pairs by hydrogen bonding will produce a 'twisting'
   as in the DNA double helix.
5 A closer approach to the real structure is obtained by bending the amino
   acid acceptor end and anticodon loop towards each other and folding the
   other two loops back so that they are almost in contact.

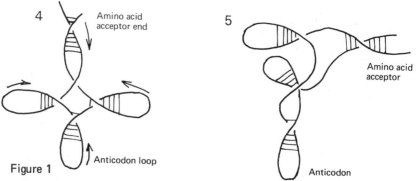

Figure 1

A perspective view of the molecule is shown in Kim *et al.* (1973).

through a right angle. Unpaired bases form loops of single-stranded RNA at the corner and closed end. The anticodon is contained in the loop at one end of the L and the amino-acid binds onto one of the prongs at the other end, these regions being some 82 Angstroms apart. It is interesting to note that one of the loops at the corner could easily be enlarged without distorting the rest of the molecule and could thus accomodate the additional nucleotides found in other tRNA molecules. Another feature is the accessibility of the amino-acid acceptor site which projects out from the rest of the molecule and this could be important in the binding and release of the amino-acid. Higher resolution data should soon be available which will allow an analysis of the interactions which, stabilise the molecule in this particular three-dimensional conformation. However, it will probably be some time yet before detailed structural information is obtained concerning other aspects of the ribosomal protein assembly plant.

While we are on the subject of nucleic acids it is worth while recording some exciting work on the antibiotic *actinomycin*. It is especially important to have accurate three-dimensional knowledge about drugs, not only so that we can find out how they act, but also so that we can make sensible guesses about how to enhance their effectiveness and reduce their toxicity. Actinomycin is a fairly complex molecule having two cyclic polypeptide regions linked through a phenoxazone ring system. It binds to the DNA double helix and specifically inhibits RNA synthesis. X-ray crystallographic studies on co-crystals of actinomycin with its DNA substrate *deoxyguanosine* show how this comes about (Sobell *et al.*, 1972). The phenoxazone ring system inter-calates into the DNA helix between adjacent base-pairs, and presumably manages to do this because of its considerable similarity to such base-pairs. The peptide chains lie in the narrow groove of the helix and interact specifically with deoxyguanosine residues on opposite chains through hydrogen bonds (fig. 2). These hold the molecule firmly in position and of course its interruption of the DNA sequence plays havoc with the RNA polymerase reaction, so that actinomycin becomes a candidate for the control of malignant cells. There are many drugs whose function

44

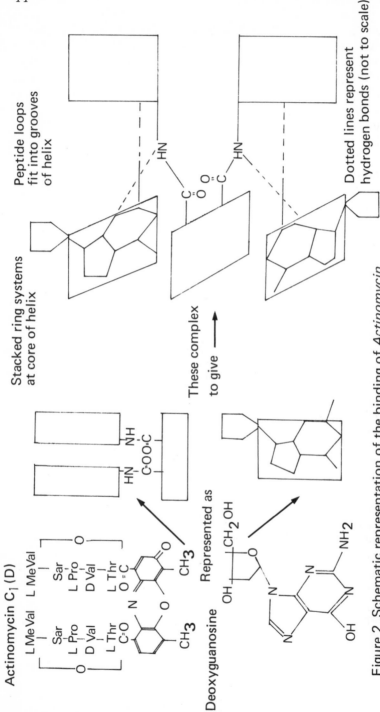

Figure 2 Schematic representation of the binding of *Actinomycin* to *Deoxyguanosine*. For more detailed diagrams and photographs see Sobell *et al.* (1972).

depends on a specific interaction with a cellular 'receptor' molecule, but actinomycin is one of the first where we have a clear and detailed understanding of the interaction mechanism. With this knowledge it is hoped that some way can be found of altering the molecule, which does not affect this interaction, but reduces the toxicity of the drug by changing its relative permeability into normal and malignant cells.

Membranes: structure and transport
An understanding of the diverse, and usually highly complex functions of cell membranes is today providing one of the greatest challenges ever to the biological scientist. Because the properties of most membranes are so intimately dependent on the spatial distribution of their constituent molecules, the problems of membrane structure are an obvious target for examination by the various biophysical techniques, and indeed practically every technique available is being thrown into the attack. From our point of view the trouble with membranes is that they are not crystalline, or at least only to a small extent, so that X-ray crystallography is of limited application. Such work as has been done has been concerned mostly with nerve myelin and we now have a fairly good idea of its transverse electron density distribution. The interpretation of this density distribution in terms of the occurrence of protein, lipid, or cholesterol is rather difficult, but at the very least the results are compatible with a bilayer structure. There has been an argument raging for some time about whether membranes are basically simple phospholipid bilayers or, as recent electron microscope studies have indicated, whether they might not be some kind of two-dimensional array of lipo-protein particles (fig. 3). The general consensus now seems to be that it depends on the membrane. Those with a low protein content and relatively simple function, like myelin, are probably bilayers similar to those originally envisaged by Danielli and Dawson (1935). As the protein content rises the structure becomes increasingly complex, with more and more protein becoming immersed in a sea of lipid, until a highly ordered particulate system is achieved, as in the inner mitochondrial membrane.

Even the bilayer is not what it used to be. Electron spin

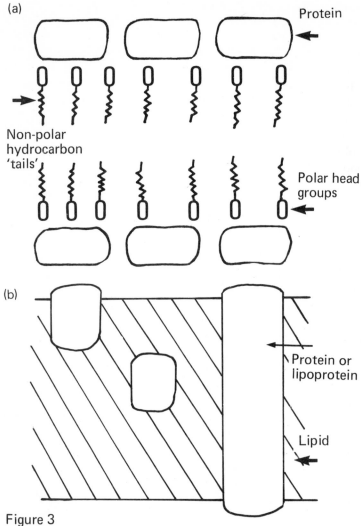

Figure 3

(a)   The phospholipid bilayer model of the cell membrane as envisaged by Danielli and Davson (1935).

(b)   Recent work suggests the protein component is more intimately involved with the non-polar parts of the phospholipid although the latter is still probably predominantly in the bilayer form.

resonance and proton magnetic resonance studies have shown that while the polar head groups of the phospholipids might be fixed in the membrane surface, the hydrocarbon tails become increasingly more mobile towards the centre of the membrane where an almost liquid-like state exists. Interestingly this mobility is greatly influenced by the addition of cholesterol, a commonly occurring steroid in cellular membranes. It is now possible to synthesise cell-like vesicles a few hundred Angstroms in diameter bounded by a single phospholipid bilayer (liposomes) or a multilayer. These provide good model systems for testing various hypotheses, especially since a number of synthetic phospholipids can be used, incorporating special groups of atoms highly sensitive to the incident radiation of the spectroscopic techniques used to investigate them. It is possible to follow certain changes in the structure of these synthetic membranes as they interact with a variety of drugs, hormones, and other small molecules known to affect naturally occurring cell membranes. However, the time has not yet arrived when we really understand the meaning of these changes.

The concept of a layer of protein sitting on the polar surface of a lipid bilayer has also come in for some criticism. Infra-red, circular dichroism, and optical rotatory dispersion studies, all of which can detect the various elements of protein secondary structure, such as the alpha helix, suggest that much of the protein has a non-polar environment and must presumably be surrounded by the lipid hydrocarbon chains. This is in accord with electron microscope studies using the freeze-fracturing process which is supposed to split the membrane down the middle. The photographs show a distinct particulate nature of the exposed (inside) region (Photo 1).

One of the most fascinating developments in the last few years has been the use of fluorescent 'probes'. These are special molecules (such as 1-anilinonapthalene-8-sulphonate) that contain atomic groupings which fluoresce after an initial period of irradiation. Such molecules can be intercalated into the membrane with their polar regions sited next to the polar head groups of the phospholipids. By using a variety of molecules, where the fluorescent part is at different distances from the polar part, it is possible to probe the membrane at

Photo 1 Freeze-etched replica of a fracture plane through plasmalemma of *Chaetomorpha Melagonium*. Granule diameter about 175Å.

different depths. Not only is the intensity of the fluorescence dependent on the mobility and environment of the molecule—which can thus be investigated—but experiments with mitochondria suggest that it is also dependent on the energy processes taking place. We can now 'watch' different parts of the mitochondrial membrane at work.

These are of course early days to expect detailed results concerning the finer points of membrane structure, but gradually a picture is emerging which must surely soon be brought into sharp focus. At the moment, for really fine detail, we must move to allied problems such as the structure of certain antibiotics and hormones which affect the transport properties of membranes. A good example is the cyclic depsipeptide antibiotic *Valinomycin*. This molecule is a 'carrier' which ferries alkali metal ions to and fro across biological membranes. Not only does it enable these charged atoms to 'dissolve' in what is essentially a non-polar barrier but it will differentiate between ions of different size. A knowledge of the mechanism by which it does this could be vital to our fuller understanding of the way cells maintain considerable ionic concentration gradients between their cytoplasm and the interstitial fluid. It turns out that the molecule is shaped like a rather fat ring doughnut with six carbonyl groups pointing towards the central hole which just happens to be the right size to take a potassium ion. The ion is held in position by a beautiful sixfold coordination with the carbonyl oxygen atoms. Moreover, the surface of the molecule is covered with non-polar amino-acid side-chains allowing easy diffusion into a lipid membrane (fig. 4, Pinkerton *et al.*, 1969). Ions which are smaller or larger can

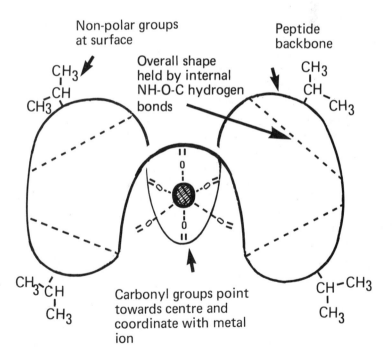

Figure 4 Schematic view of *Valinomycin*

still be bound, though not so efficiently, and the antibiotic will discriminate against them in favour of K $^+$.

Other molecules such as *Gramicidin A* and *Alamethicin* would appear to induce pores in the membrane which allow the passage of hydrophilic substances. Alamethicin is particularly interesting since the rate of transport changes as a function of an applied electric potential difference across the membrane. This could have obvious repercussions in our understanding of nerve impulse transmission.

Myoglobin starts an avalanche
When the three-dimensional structures of myoglobin and haemoglobin were announced some thirteen years ago, the laborious and inventive efforts at their solution were considered worthy of a Nobel Prize. Nowadays the response to a new protein structure is more likely to be a flutter of excitement among the specialists, but this does not lessen the importance of each new discovery. At the time of writing, about thirty proteins, including sixteen enzymes, have had their three-dimensional structures determined, but probably another hundred or so are being actively investigated throughout the world and it now seems as though a new structure appears in the literature about once a month.

Enzymes and other proteins are of course vital to all living processes and an understanding of their mode of action is perhaps the most intensely studied aspect of biology today. Industry as well as Academy has an interest in this accumulation of knowledge. Enzymes as catalysts are a million times more effective than any product of man's technology. Synthetic proteinaceous food will, alas, be essential to feed our exploding population. Protein malfunction is the cause of so much disease which must be understood before it can be treated. The list is endless, but one thing is already clear: it is not enough to know just the amino-acid sequence, or the shape, or the folding pattern of the polypeptide backbone. When a protein acts we want to know which atoms move, where they move to and why; what electronic transitions are taking place and how these affect the functioning molecule.

The time-scale of this sort of investigation is measured in years and such detailed information is available in only a few cases. Eminently reasonable explanations of the mechanistic processes involved have been proposed for lysozyme, ribonuclease, carboxypeptidase, and chymotrypsin, but perhaps the most fascinating story has been the gradual unravelling by Max Perutz of the complex intricacies of haemoglobin.

In its reversible combination with oxygen the haemoglobin molecule undergoes some kind of conformational change. The Bohr Effect (a variation in oxygen affinity with pH) tells us that conformations of different binding affinities are more stable in some environments than in others. The situation is complicated by the fact that haemoglobin is a tetramer, composed of two pairs of chemically identical sub-units—the alpha chains and beta chains—so that there is the possibility of changes in both the tertiary (intra subunit) and quarternary (inter subunit) structures. Thus whereas myoglobin has a normal hyperbolic oxygen binding curve, haemoglobin exhibits a sigmoid curve and this implies that the binding of oxygen to one of the subunits so affects the overall conformation of the molecule that its affinity towards additional oxygen molecules is altered. A rather over-simplified picture (believe it or not) of the events taking place is as follows:

Because the haem pocket in the deoxy form of the molecule is slightly larger in the alpha chains than in the beta chains, it is likely that the first oxygen atom binds to the iron atom in the haem group of an alpha chain. This binding alters the electronic state of the iron atom which results in a *change in size* of the iron atom. What is happening here is that the five d orbital electron energy levels of the ferrous ion are initially split on coordination with the porphyrin ring system into a group of three and a group of two. The additional binding of oxygen causes a change from a state where there are unpaired electrons in both groups of energy levels (high spin paramagnetic state) to a state where all the six d electrons are paired in the group of three lower energy levels (low spin diamagnetic state). The change in size of the iron atom causes it to shift its position relative to the porphyrin ring and in doing so it pulls an attached histidine residue with it (fig. 5). The histidine residue forms part of one of the eight alpha helices of the sub-unit and this helix now moves closer to an adjacent helix, causing the expulsion of the penultimate

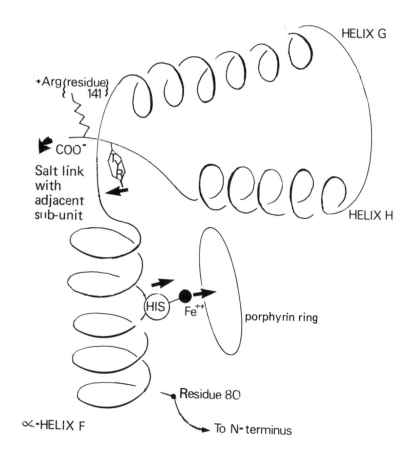

Figure 5 Schematic diagram showing major movements within haemoglobin sub-unit on oxygen binding. For additional diagrams showing consecutive binding to all four sub-units see Perutz (1970).

tyrosine residue originally contained in a pocket between these two helices. The expelled tyrosine pulls the C-terminal residue with it, so rupturing salt links which hold the subunit to its neighbours. With the removal of these constraints the molecule undergoes a change in its quartenary structure and the sub-units slide past each other to take up a more stable position. The binding of additional oxygen atoms follows a similar pattern, but because of these quarternary structure changes the affinity for oxygen is different each time.

What a marvellous delicately balanced process this is. No wonder that minute mutations of the molecule can have such catastrophic effects as sickle cell anaemia.

Teaching biophysics in universities

Undergraduate courses in biophysics are available in about twenty British universities and some polytechnics, although they may be variously disguised under such alternative headings as molecular biology, physical biochemistry or biophysical chemistry. The length of these courses varies from about twenty lectures to a three or four year full-time scheme of study. The shorter courses are usually concerned with some particular aspect of biophysics which is closely related to the main subject of study. For example, a course on nerve, muscle, and membrane biophysics and the physics of sensory perception might be included in a degree in physiology or zoology. Biochemistry departments would usually have a course describing the structure and function of biological macromolecules in physico-chemical terms. Medical schools will often present a number of lectures on such topics as the properties, uses, and effects of ionising radiations and on computing and medical electronics. Sometimes the pure science degree schemes will include a biophysics option—a third year physics student for example, could be introduced to nerve excitation phenomena, visual and audio-reception or diffraction and spectroscopic studies of macromolecules. These short courses are extremely valuable, for not only do they deal descriptively with important phenomena, they also introduce the biology or medical student to a quantitative rather than a qualitative way of thinking.

It may be that the presentation of these specialist courses is the most useful part a biophysicist can play in undergraduate education. Indeed, many people feel that more extensive courses are better suited to postgraduate study and that enough sub-divisions have already been created in the

Biological Sciences at the undergraduate level. There is no doubt an element of truth in this and certainly these specialist lectures will form the greatest part of the biophysics contribution for some time to come.

There are a few establishments (including the University of Leeds) which offer biophysics as a single subject degree scheme. The motivation here is more than just providing an exceptionally broad-based education. It is the combination of the rigorous approach of the physical sciences with an appreciation of the complexity and beauty of living systems which is felt to be so appropriate to the modern world. There is a dire need in both industry and government for graduates who can communicate in depth with both physicists and biologists. Such a degree naturally provides an ideal grounding for the potential teacher, especially in view of current trends to broaden the A-level curricula.

It is not an easy course to teach, for one must always bear in mind the mathematical limitations of the average biology student, and the shallow acquaintance of most physics and chemistry students with biology. Because acceptable students provide such a diversity of backgrounds, it is necessary for the preliminary year(s) to be concerned with attaining a certain standard in each of the subjects, mathematics, physics, chemistry, biochemistry, and biology, which is appropriate to the more advanced parts of the course. This might entail an almost individually plotted programme of lectures for each student during this initial period. The latter part of the course would deal on the one hand with modern physical and physico-chemical techniques and their application to relevant biological problems, and on the other hand to a more advanced treatment of biological structures and phenomena. Where the emphasis lies will depend to a large extent on the department concerned, even on the individual lecturer, and perhaps that is how it should be.

Teaching Biophysics in Schools

Biophysics *per se* cannot really be taught in schools. Without some background of solid state physics it will probably fail to make any impact and the study of biophysics itself is best kept to university years. A better approach would be to build the biophysics informally into biology classes by introducing physical and molecular ideas as biological problems come to be discussed. In this way also certain attitudes of mind can be fostered and encouraged which will enable the school-leaver to more readily appreciate future developments in biology, which are expected to become increasingly physical in content. Some suggestions about topics which lend themselves to the biophysical approach are given below, but any move which lessens the terrible dichotomy which exists between biology and mathematics/physics is to be welcomed.

1 *Population studies*

The wide variability between species, and between individuals within a species, can be discussed in terms of the variability of the basic structural material. Only proteins have the required variability—the existence of twenty different amino-acids means that a protein containing only a hundred amino-acids could occur in 20^{100} different chemical forms. At the same time the variability must be conservative in the sense that any member of a species retains the species characteristics. There must therefore be informational molecules which prescribe the protein structure and which are themselves protected against variation. The differences between the species are reflected in differences in three-dimensional structure of the constituent proteins and perhaps some of these could be discussed in a simple way. Such three-dimensional differences are not required for the informational molecules since changes in a one-dimensional linear assembly of code-words will suffice to produce three-dimensional changes in the proteins.

2 *Membranes*

The tendency for surfaces to assume states of minimal energy and therefore minimal area could lead directly to the structure of a lipid monolayer on water and hence to a lipid bilayer around the organelles in a cell. The idea of electric charges on membranes could perhaps also be introduced in terms of the charges on colloid particles which can be demonstrated.

3 *Cell physiology*

Spectroscopy, at least optical spectroscopy, could be brought in here in terms of chlorophyll and photosynthesis. It might even be pointed out that the energy required to convert one molecule of carbon dioxide to glucose needs at least four quanta (which would need to be explained) and these four must be collected at one place over an appropriate short time interval. This means that energy must be transferred from one place to another closely neighbouring place to get the energy of the four quanta together, and this leads to the idea of energy transfer along molecules.

4 *Anatomy*

Nature has utilised the progressive loss of solubility and reactivity which occurs on condensation of small, highly reactive molecules (amino-acids, sugars) to produce long linear polymers whose parallel and/or ¹aminar aggregation gives rise to structures which are extraordinarily strong relative to their density. This has enabled large complex living creatures to have formed. The properties of such diverse supporting structures as skin, bone, teeth, hairs and even cell walls are clearly related to the underlying molecular arrangement. Perhaps this is a stage where molecular models could be introduced. The molecular structures of hair, silk, collagen and muscle for example, are fairly easy to represent even if a trifle laborious. Commercial models are becoming more readily available and the new plastic versions are getting cheaper every year. Stiff wire and polystyrene can also be used to good effect. The model forms the base camp of the exploration. It has to be built with a knowledge of atomic structures, chemical bonds and electrostatic interactions and leads, for example, to the interdigitating actin and myosin microfibrils of muscle and the explanation of the contractile process.

5 *Whole organism physiology*

The necessity for gas exchange (diffusion laws!) and the intake of food across the surface of an organism limits the size of an organism without specialised systems. The pure physics of these processes has led to the diversity of feeding and gas transfer systems. Examples here would be leaf form in plants and lung development in animals.

6 *Growth*

The present awareness and concern, particularly among the

young, of the delicate balance between man and his environment provides an ideal opportunity to introduce the more simple mathematical laws which describe the population explosion and resource depletion. The rates of growth of various organisms, and the factors which influence them, provide endless examples to aid the teaching of calculus. The ubiquitous exponential function probably provides the strongest quantitative link between biology and physics.

References

J. F. Danielli and H. Dawson (1935) *Journal of Cellular and Comparative Physiology*, 5.

R. E. Dickerson, T. Takano, D. Eisenberg, O. B. Kallai, L. Samson, A. Cooper and E. Margoliash (1971) *Journal of Biological Chemistry*, 246.

S. H. Kim, G. T. Quigley, F. L. Suddath, A. McPherson, D. Sneden, J. J. Kim, J. Weinzierl and A. Rich (1973) *Science*, 179.

M. Perutz, (1970) *Nature*, 228.

M. Pinkerton, L. K. Steinrauf and P. Dawkins (1969) *Biochemistry & Biophysics Research Commun*, 35.

H. M. Sobell, S. C. Jain, T. D. Sakore and G. Ponticello (1972) *Cold Spring Harbour Symposia on Quantitative Biology*, 36.

J. Takano, R. Swanson, O. B. Kallai and R. E. Dickerson (1972) *Cold Spring Harbour Symposia on Quantitative Biology*, 36.

Further Reading

Easier reading

L. E. Orgel (1973) *The Origins of Life*, Chapman and Hall.

R. E. Dickerson and Geis (1969) *The Structure and Action of Proteins*, Harper and Row.

James D. Watson (1971) *The Double Helix*, Penguin.

A. P. M. Lockwood (1971) 'The Membranes of Animal Cells', *Studies in Biology*, 27, Edward Arnold.

M. Tribe and P. Whittaker (1972) 'Chloroplasts and Mitochondria', *Studies in Biology*, 31, Edward Arnold.

More difficult reading

R. Buvet and C. Ponnamperume (eds), (1971) *Molecular Evolution, 1. Chemical Evolution and the Origin of Life*, North Holland.

C. C. F. Blake (1972) 'Progress in Biophysics and Molecular Biology', vol. 25, *X-Ray Studies of Crystalline Proteins*, Pergamon Press.

H. R. Wilson (1966) *Diffraction of X-Rays by Proteins, Nucleic Acids and Viruses*, Edward Arnold.

D. Chapman (1968) *Biological Membranes*, Academic Press.

Developmental biology

MICHAEL BALLS

Development is the process whereby a single cell (the zygote) gives rise to the different cell types that function in a coordinated way in a multicellular organism. The cell is the fundamental unit in development, and interactions within and between cells lead through successive cell divisions to cells with very different phenotypes. Developmental biology, stimulated by the remarkable progress in molecular biology of the last fifteen years, has now become a central theme in biological research, linking molecular biology with studies at tissue, organ and organism levels. This brief review is a highly selective and personal survey. The field is advancing at such a rate and in so many directions that it defies more general analysis within the space available.

Differentiation and differential gene action

Phenotypic differences between cells are due to differences in structural proteins and enzymes which, in turn, are due to differential activation of the genome (i.e. genetic content of the cell). Cell differentiation involves differential *determination* (selection of particular genes for future activation and expression), followed by differential *gene function*. Determination and overt differentiation may be separated by many cell generations, as in insect imaginal discs (see Hadorn, 1968). The term *cell differentiation* is normally restricted to the process which leads to *stable* differences between cell types (e.g. those which distinguish hepatocytes from granulocytes), as distinct from reversible *modulations* (e.g. differences between hepatocytes synthesising or breaking down glycogen). Much attention is currently being paid to the control of cell differentiation, and many levels of control have been identified (see Markert & Ursprung, 1971):

1 *Variation in DNA content*, by (i) *Polyploidization*—
   multiplication of the entire somatic cell chromosomal
   complement, as in mammalian hepatocytes; (ii)
   polyteny—multiple replication of DNA, where the
   strands remain attached side by side, producing giant
   chromosomes, as in dipteran larvae; (iii) *gene
   amplification*—an increase in the numbers of copies of
   specific genes, such as ribosomal RNA genes in
   amphibian oocytes. The polytene chromosomes
   of *Rhynchosciara* and *Sciara* also show gene
   amplification—different parts of the genome are replicated to different degrees, resulting in visible DNA
   puffs.

2 *Differential transcription of DNA* occurs, since different
   cells contain different RNAs (shown by competitive
   DNA-RNA hybridization—see Markert & Ursprung,
   p. 23); different RNAs are synthesised at different rates
   at different stages of development; RNA puffs on
   dipteran polytene chromosomes vary in position according to tissue and developmental stage, and genetic
   studies link individual puffs with particular proteins.

3 *RNA transport and processing*. Not all the RNA
   transcribed in the nucleus is transported to ribosomes in
   the cytoplasm—much of it is quickly broken down in
   the nucleus. There is selective stabilization and transport
   of particular RNAs.

4 *Control of translation*. Long intervals between transcription and translation of the message into protein occur
   both in embryogenesis and in cell differentiation (e.g. in
   erythropoiesis and lens formation). Long-lived RNA
   appears to be associated with protein (as 'informo-
   somes'). Differential timing of initiation and rate of
   translation give further levels of control.

5 *Control of protein assembly* and epigenetic modification
   of proteins occur. Lactate dehydrogenase exists in five
   forms (isozymes) and is composed of tetramers of two
   types of sub-unit, coded for by different genes. The
   proportions of the various isozymes vary in different
   tissues and in the same tissue at different stages of
   development—according to the availability of sub-units.
   Deletion of a part of trypsinogen gives the functional

molecule—trypsin. Different forms of glycogen phosphorylase based on conformational variation have different specific activities, and interconversion among the various forms regulates the total activity of the enzyme in the cell.

Oogenesis and early development

In most animals, *oogenesis* is a prolonged period (usually a protracted meiotic prophase) which, in addition to providing a haploid female gamete, is the basis of *embryogenesis* (the formation of the embryo) through growth, intensive biosynthesis and the organisation of differentiated cytoplasm. Development begins, not at the moment of fertilisation, but early in *gametogenesis*. For example, the amphibian embryo contains energy (in the form of yolk) sufficient to produce a swimming tadpole, long-lived informational molecules which contribute to early development, and enough ribosomes and mitochondria for 60,000 to 100,000 cells. This allows the non-motile, unresponsive zygote to develop very rapidly into an active, swimming tadpole. In addition, the organisation of the cytoplasm produced in oogenesis is the basis of the organisation of the embryo.

The animal egg is a dynamic system, activated by the male gamete which also restores the diploid chromosome number. There follows a period of rapid DNA synthesis and cell division (*cleavage*) without the need for synthesis of cytoplasm, since the organelles, enzymes and energy synthesised and stored during oogenesis are distributed among the cells, and the cell generation time is reduced to thirty minutes (the fastest time recorded for adult *Xenopus* cells *in vitro* is fifteen hours). The formation of the embryo is based not only on the activity of the oocyte genome, but also on cooperation between the oocyte and the somatic cells of the mother. For example, yolk proteins are made in the liver and transported via the blood plasma and the follicle cells to the amphibian oocyte, where they are assembled into platelets (Redshaw, 1972); the precise organisation of the mollusc egg cytoplasm is influenced by the surrounding follicle cells (Ubbels *et al.*, 1969; Raven, 1970).

Nucleo-cytoplasmic interactions

Although the early embryologists realised the significance of special regions in the egg cytoplasm, and the importance of cleavage in producing many genetically identical nuclei whose genes could be differentially activated by different cytoplasmic components, it has been difficult to link particular cytoplasmic components with particular developmental events. Two techniques are currently proving of great value in investigating nucleo-cytoplasmic interactions—nuclear transplantation and cell hybridisation.

Every naturally occurring change in gene activity (so far investigated by nuclear transplantation into amphibian eggs or oocytes) can be induced or repressed by normal cytoplasm constituents. Adult nuclei inserted into activated unfertilised *Xenopus* eggs are frequently induced to behave as zygote nuclei, and can give rise to fertile adults (Gurdon, 1968). This shows that the genes not activated during the development of a particular cell type are not lost, but are retained in an inactive form, and that the whole genome can be reactivated when the nucleus is placed in egg cytoplasm. Nuclei transferred to growing oocyte cytoplasm take on the characteristics of growing oocyte nuclei (large volume, multiple nucleoli, no DNA synthesis, high level of RNA synthesis). Hybrids (produced by cell fusion) induced by adding inactivated Sendai virus (which promotes cell fusion) to co-incubated cell lines *in vitro* also indicate cytoplasmic control of nuclear activity. The classic example is the reactivation of the hen erythrocyte nucleus by HeLa cell cytoplasm (Harris, 1970). Evidence from both these techniques shows that the selective uptake of protein by the nucleus is the basis of cytoplasmic control of nuclear activity.

Embryogenesis and Organogenesis

During *gastrulation* the cells produced by cleavage are reorganised to form the three-layered embryo. Concomitantly, two other events of major significance also occur. Firstly, information produced by the new genome becomes necessary for further development; for example, paternal effects become apparent, interspecific hybrids stop developing. Secondly, the fates of the main cell types are determined as a result of nucleocytoplasmic interaction and cell to cell interaction (embryonic induction). *Epithelio-mesenchymal* interactions are the basis of morphogenesis in many organs, such as kidney, pancreas, salivary glands (Fleischmajer & Billingham, 1968), but their molecular basis is not understood. Such interactions provide an integration of different development pathways, so that each cell type is formed at the correct moment and in the correct position. The genome must be involved both (i) in controlling the release of morphogenetic signals and (ii) in controlling the ability of cells to respond and the nature of the response. One must also be aware of the distinction between individual cell differentiation and differentiation in groups of cells (known as *fields*). These two processes are closely related and interdependent, but not necessarily identical, phenomena.

Programmed *cell death* is an important component of cell differentiation which occurs during the normal development of many organs. For example, cell death is essential for the shaping of the vertebrate limb, the formation of fingers and toes, and the opening of eyelids. In metamorphosis of amphibian larvae under the influence of thyroxine, some cells die, whereas others proliferate and begin to express their cell type-specific function (Tata, 1973). Similarly, in the development of sex organs from the sexually undifferentiated gonad, which contains rudiments of both sexes, male or female hormones lead to the death of one cell type and proliferation and differentiation of the other. The response of a particular tissue is thus highly dependent on its state of differentiation.

The term *pattern* is used to describe the non-random occurrence of cell types, or their products, to give an arrangement that suggests design or orderly distribution: for example, the arrangement of cells in the liver or pancreas, patterns in bird feathers, the arrangement of pigment cells in amphibians, bristle patterns in insects, the arrangement of the scales and hinges of butterfly wings, and the distribution of stomata on plant leaves. Unconvinced by the dogma that pattern (like all biological phenomena) is totally explicable in terms of molecular differentiation, Wolpert (1973) has proposed that a different set of general rules and principles, acting at the *cellular* level, apply to the expression of genetic information in terms of pattern and form. He has introduced the concept of *positional information* to provide a conceptual framework for use in analysing pattern formation and its regulation. The concept involves a two-step process: the assignment of positional information to the cells in a coordinate system, then the interpretation of this information by the cells and the expression of differentiation according to their genetic make-up and developmental history. This concept is currently being used in studies on regeneration in *Hydra*, the organisation of the insect integument, limb formation in vertebrates, and amphibian and sea urchin embryogenesis. This concept is attractive, partly because it provides a mechanism whereby small, simple molecules (hormones, inducers, morphogens) could have profound yet different effects on different kinds of cells.

The cell cycle and development
The cell cycle (the interval between the formation of the cell by division of its mother cell and its division into two daughter cells) is a fundamental unit of time at the cellular level (Mitchison, 1971). However, its phases are ignored in almost all the text books currently used in biology teaching (Balls and Godsell, 1973), which frequently equate cell division with the four-stage mitotic phase (prophase to telophase), following the interphase, that period of intense synthetic activity so often misnamed the 'resting stage'. It is becoming clear that controls acting in the presynthetic phase (G1), during DNA synthesis (the S phase), in the postsynthetic phase (G2) and during mitosis itself, are of great importance in normal development and differentiation (Balls and Billet, 1973), and in the development of cancer (Baserga, 1971). Passage through the S-phase and mitosis appear to be essential precursors to major changes in cell type-specific activity (Malamud, 1971). Each daughter cell is probably programmed like its parent cell unless developmentally important agents enter during certain periods in the cell cycle. It is known that division is a pre-requisite for changes in determination ('transdetermination') in insect imaginal discs (Hadorn, 1968). Lawrence *et al.* (1972) had to add cell division as an essential factor in pattern change when they attempted to reconcile experimental data with the positional information theory. Gurdon (1973) proposes that unequal distribution of cytoplasm during cell division can lead, for

example, to differentiation and specialisation of one daughter cell while the other remains undifferentiated. Position and contact with other cells can also result in different fates for the daughter cells produced by cell division. For example, after division in the vertebrate epidermis, the cell remaining in contact with the basement membrane divides again, while the other daughter cell differentiates, functions by producing keratin, eventually dies and is sloughed off. Similar zones of proliferation, differentiation, function and loss (according to position) are found in the villus of the intestine.

The reversibility of differentiation

Is the sequential, selective programme of gene activation and repression which leads to differentiation reversible? Can a new developmental programme be undertaken once a cell has reached a seemingly terminal state of differentiation? These are controversial questions and some still doubt that cells of one type can transform to become cells of a different type. I consider that such changes can, and do, occur under natural and experimental conditions though cell proliferation is a necessary part of the process. A single carrot phloem cell, cultured in a nutrient medium based on coconut milk with appropriate hormones added, will develop into a complete carrot plant (Steward, 1970). This differs from Gurdon's experiments with adult amphibian nuclei, which involved the reversal of *nuclear* differentiation. The best examples of cell reprogramming in adult animals occur in the regeneration of lost parts.

1 *Amphibian lens regeneration*

In many salamander species, a new lens will form from the iris after lens removal. This regeneration requires the presence of a lens-inducing factor from the retina and occurs in three main stages. Firstly, a *depigmentation* stage, when the iris cells dedifferentiate by discharging their pigments. Secondly, a *multiplication* stage, when the depigmented cells show active DNA synthesis and cell proliferation. An increase in RNA synthesis precedes the third stage, *fibre differentiation*, when nuclei disappear and the α, β and γ crystallins are formed. Since the original lens was formed from epidermis and the iris from neural tissue, lens formation from iris cells represents transformation between two cell types that are functionally, morphologically and embryologically distinct.

2 *Amphibian limb regeneration*

When the limb or tail of an adult newt is amputated, the epidermis seals off the wound, then muscle, cartilage, bone and connective tissue cells dedifferentiate and form the blastema. This reserve of undifferentiated mesenchyme cells proliferates and then redifferentiates into the muscle, cartilage, bone and connective tissue cells of the new limb. Epidermis, nerves and blood vessels grow out from the intact tissues in the stump and do not develop from blastema cells. Weiss (1973) considers that this represents *modulation* rather than *redifferentiation*, but I cannot accept the idea that dermis, cartilage, bone, sarcolemma and nerve sheath cells are all the same type of cell responding differently to different environmental conditions (modulation). Because we cannot yet provide the conditions required for reprogramming a particular cell type into *all* other cell types, this does not mean that such reprogramming cannot and does not occur.

3 *Mammalian liver regeneration*

If a part of the liver of a mammal is removed, rapid cell proliferation occurs in the remaining part. Although there is little structural evidence of dedifferentiation, biochemical evidence (from competitive DNA-RNA hybridization—see Markert and Ursprung, 1971) shows that the new cells pass through a programmed sequence of gene activities similar to those occurring during normal liver development.

The control of cell, tissue and organ growth

Although adult mammals reach a finite size and appear to stop growing, the processes of differentiation, morphogenesis and growth continue at the cell and tissue level. Genetic and environmental factors exert an influence, but there is also close control by the organism of the sizes of the various tissues and organs relative to the total size of the organism. There is great variation in the rates of cell turnover in adult tissues, from those in which cells divide rarely if at all (striated muscle, nerve, brain) to those with a high rate of tissue renewal (epidermis, intestinal epithelium, blood-forming tissues). The cells of some organs which normally have very low rates of cell proliferation will divide if a part of the organ is removed. For example, if one kidney is removed, the remaining kidney increases in size by a process known as *compensatory hypertrophy*. If two rats are connected so that there is vascular continuity between them (parabiosis), and part of the liver or one of the kidneys is removed from one animal, cell proliferation increases dramatically in the liver or kidney of the other animal, as well as in the tissue remaining in the operated animal. Just as many metabolic processes are controlled by hormones carried in the circulation, these experiments suggest that the stimulation and inhibition of hypertrophic growth are forms of *humoral regulation*. Many explanations have been suggested, but the most satisfactory concept is that of Weiss and Kavenau (1957) who suggested that growth and differentiation are controlled by tissue-specific stimulators and inhibitors produced by cells of the same type. The problem with this concept, as with the concept of embryonic induction, has been to identify the substances involved.

As was mentioned earlier, tissues with a high rate of cell turnover show a steady state relationship between cell proliferation, differentiation, function and loss. The disturbance of such balanced, dynamic systems can have serious consequences for the organism. For example, radiation damage of the proliferative zone of the gastrointestinal epithelium can lead to intestinal bleeding, similar damage to that of the epidermis can produce cracked skin; and an excess of cell production over cell loss can lead to tumour formation. Although many books describe cancer in terms of 'the outcome of uncontrolled and haphazard cell multiplication', 'tumours with unlimited growth' and 'cells locked in the mitotic phase', cell differentiation, function and cell death *do* occur in tumours, as in normal tissues. Cell division and growth in cancer tissue are *not* uncontrolled (many malignant tumours have a lower mitotic incidence than some of the normal tissues of the body). The problem results from alterations in the relationships between the parts of the system, for example, because of a shortening of the total cell cycle time, an increase in the number of cells in the population which are engaged in cell proliferation, or a fall in the rate of cell loss. Thus, in the long term, an approach to the cancer problem based on the identification of the factors promoting specific cell differentiation might be very rewarding. For example, some forms of leukaemia are characterised by large numbers of immature leucocytes, but the negative feed back mechanisms that control production from stem cells might be based only on substances produced by mature cells. If the conditions which promote differentiation from stem cell to mature cell could be identified, perhaps the immature cells of leukaemia patients could be persuaded to mature, age and die. This would be more satisfactory than the empirical treatment methods used at present, or those based on inhibiting mitosis in general, which have side effects on organs where cell proliferation is a requirement for the maintenance of normal structure and function.

Another interesting approach to cancer, as an abnormality of cell differentiation, is based on the suggestion that the interactions between cells and tissues which lead to the development of organs (*morphogenesis*) might also be responsible for their normal maintenance (*morphostasis*), and that disturbance of these mechanisms might result in neoplastic (cancer) development—*carcinogenesis* (Tarin, 1972 a,b). Although a number of agents (viruses, radiation, chemicals) cause cancer, their mode of action is not understood. Much attention is being paid to cell transformation at the molecular level, but, just as a multicellular organism cannot be built up from *parts* of cells, I doubt whether cancer will be fully understood if we concentrate on the subcellular level. Cancer should be seen as a multistage process involving a sequence of developments—transformation of a cell or cells; alterations in the normal relationships between cells and tissues leading to a disturbance in the normal relationship between cell proliferation and cell loss; and failure of the organism's surveillance mechanisms to deal with the abnormal cells.

Although some cells in monolayer culture retain their type-specific function for long periods *in vitro* (especially endocrine gland-derived tumour cells producing hormones), most cells in culture lose their specific functions and produce aneuploid cell lines unlike any of the cell types found in the body. Such cell strains result from selection for survival and proliferation *in vitro*, and resemble bacterial cultures in that their response to different culture conditions is more comparable to *modulation* than to cell differentiation. However, this is not the case with organ culture, where three-dimensional relationships between cells are retained, where normal differentiation *in vitro* occurs, and where the normal cell type-specific structure and function can be maintained for extended periods (Monnickendam and Balls, 1973). Organ cultures have great potential value for future studies on the maintenance of normal function, on functional failure and on carcinogenesis.

Pleiotypic effects and congenital abnormalities

Humans are subject to a frightening incidence of congenital abnormalities (in one per cent of births), which affect the development of almost all organs. Mutants are known in animals (particularly mice), which cause abnormalities similar to those occurring in man. A single Mendelian unit is often responsible for a complex syndrome, i.e. single genes have *pleiotypic effects*. For example, congenital hydrocephalus in the mouse is a cartilage anomaly which has secondary effects on the entire skeletal system, and tertiary effects on many other systems contained within or attached to it. Other factors affecting development also have pleiotypic effects, e.g. the drug thalidomide, German measles virus, and hormones. Simulations of human congenital defects can be produced in animals by physical or chemical treatments. Animal mutants and simulated diseases are useful in investigating the primary causes of abnormalities in man and in their treatment. The injection of pituitary hormones into the pituitary dwarf mouse (*dw*) corrects some or all of the secondary and tertiary symptoms resulting from the mutation. Similarly, daily administration of L-thyroxine to human cretins raises their IQ, if treatment begins on the first day of life. In relieving the symptoms of congenital defects, a necessary and entirely moral act, we are increasing the burden of congenital abnormality carried by the human race as a whole — particularly if the abnormality has a genetic basis and the individual bearing it is able to breed. That, however, is a dilemma beyond the scope of this brief review.

TEACHING AND RESEARCH IN DEVELOPMENTAL BIOLOGY IN UNIVERSITIES

Up to now, the traditional approach to biology teaching has been *subject-centred*, and biochemistry, biophysics, microbiology, genetics, botany and zoology are separate disciplines (often taught in separate departments housed in separate buildings). I consider that this approach should be supplemented (or better, replaced) by a multidisciplinary *problem-centred* approach to important biological questions. It is biological problems that bring students and scientists together, and persuade the general public to agree to support biological research from public funds. These problems include birth control, congenital defects, organ malfunction, disease and immunity, cancer, ageing, animal behaviour, and population biology. It is vital that university biologists continue to stress the importance of the link between teaching and research, so the following observations apply to both these inseparable aspects of the university's function.

The following trends should be seriously considered in planning future approaches to developmental biology in universities:

1 *Mammalian development* (including human development) has recently become more amenable to study because of dramatic technical advances, and rapid progress is being made on many fronts. For example: allophenic mice (genetic mosaics arising through the combination of blastomeres from embryos of different genotypes) are providing much useful information about striated muscle differentiation, melanocyte differentiation and distribution, immunological tolerance, cancer and the genetic basis of behaviour; there is evidence that early mammalian development differs significantly from that of other animals, and that these differences are related to viviparity (Woodland and Graham, 1969); the early mouse blastomeres are equivalent and differentiation into the two main early cell types, inner cell mass (which forms the embryo proper) and trophoblast (which implants into the uterus wall and forms the embryonic part of the placenta), is based on positional information ('inside-outside differentiation', Graham, 1971, 1972); the maternal-foetal relationship involves modifications in the immunological

response, so that the mother does not reject the foetus as she would a skin graft. These and other recent events have two important implications. Firstly, the problems of human birth control, infertility, congenital abnormality and of mammalian development in general would be better approached by concentrating on mammalian development itself, rather than by the continued study of sea urchins, molluscs, amphibians and birds, whose use in experimental embryology is more traditional. Secondly, a moral stand must be taken on what kinds of experiments are appropriate and acceptable to society. Fertilisation and early development can be achieved *in vitro* and the blastocyst transferred to the uterus where it will implant and develop. The day of the test-tube/incubator baby is still some way off, but the day of the allophenic human is not.

2 *Plant developmental biology* has recently emerged from a long period of descriptive and physiological bias into an era where the questions being asked are similar to those concerning animal developmental biologists. Interesting developments include: techniques for plant cell and embryo culture *in vitro* which have immense implications for agriculture and plant genetics; intriguing studies on morphogenesis in fungi such as *Blastocladiella* (Truesdell and Cantino, 1971); and studies on the role of the cell cycle in differential growth in root and shoot meristems (see Balls and Billett, 1973). Courses on developmental biology should include both the common and the distinct elements of plant and animal development.

3 *Problems of multicellular organisms* should be studied in multicellular organisms. Although microbiologists talk of development, differentiation and morphogensis in bacteria, it is debatable whether these are the same phenomena as those occurring in multicellular organisms. Weiss (1973) argues that they are modulations, and that the concept of cell strain-specific differentiation should be reserved for organisms which show a separation of somatic and reproductive cells. Developmental biologists have now learned to resist the temptation to reinterpret their old data in terms of the Jacob-Monod hypothesis on the regulation of gene action in bacteria (see Markert and Ursprung, 1971). They are not so

consistently subjected to the argument that *E. coli* is the most suitable organism for studies on the control of cell division, cell function and cancer in multicellular organisms, and are more sceptical about the dogma that the order of the amino-acids in proteins is the ultimate explanation of all biological phenomena, and thus the logical place to study them. Consequently, there is a common feeling that cellular and supra-cellular levels of control also exist. However, I consider that an advanced course on *morphogenesis* (including consideration of viruses, bacteria, fungi, plants, invertebrates, vertebrates and man) is essential for any forward-looking biology department.

THE TEACHING OF DEVELOPMENTAL BIOLOGY IN SCHOOLS

It appears, from my own experience and from discussions with many undergraduates, that developmental biology is a disaster area in secondary schools. It seems that this subject is either omitted from the syllabus or course altogether, or taught on an entirely descriptive basis. For example, I am amazed that Grove and Newell's (1942) comparison of the embryology of *Amphioxus,* the frog, the chick and the rabbit is still being used in 1973 (how many of the many generations of biology students ever hear of *Amphioxus* again?). Words such as 'deutoplasm' (= yolk), long discussions of morphogenetic movements (whilst omitting other gastrulation events), slides and illustrations of saggital, longitudinal and transverse sections, three-dimensional drawings and models, whole mounts and stage-by-stage descriptions of events destroy interest in the subject while it is in its infancy. Development is not made up of stages, but is a continuous, integrated and above all, *dynamic* process. Much of what is happening in *Xenopus* or *Rana* eggs can be observed in detail as it happens, with the aid of a stereomicroscope. These events have a molecular and cellular basis, about which much is now known. More recent text books are little better. For example, Roberts (1971) more or less ignored development in what, I hasten to add, was an otherwise quite outstanding book.

I can suggest three ways of alleviating these problems:

1 *The purchase of books for the school library.* Teachers should allow the usual text books neither to influence their teaching of development nor their students' attitudes towards the subject. Instead, they should encourage wider reading, and I suggest the following:
   (i) Ebert and Sussex (1970) *Interacting systems in development*
   (ii) Markert and Ursprung (1971) *Developmental genetics*
   (iii) Spratt (1971) *Developmental biology*
   (iv) *Oxford Biology Readers* numbers 16, 25, 26, 44, 46, 51, 70, 75, 79.
2 *The use of films about development.* A number of films on various topics are available for hire (see Hinchliffe, 1972).
3 *Improve practical classes in developmental biology.* One of the main problems with this subject at the school level is mounting practical classes which are sufficiently closely related to modern developments to be interesting to students. The *Nuffield Advanced Science laboratory guide* (Sands, 1973) is a brave attempt, and many of the experiments in the *Laboratory Manual of Cell Biology* (Hall and Hawkins, 1973) would be appropriate.

The subject is best introduced as an aspect of *human biology*, concerning the problems mentioned earlier (birth control, infertility, congenital abnormalities, functional failure, cancer) and of *plant biology*, as part of the green revolution in the breeding and growth of better food plants. The *problem-centred* approach to developmental biology teaching in schools makes the subject interesting, not only for the students, but also for the teacher.

References

M. Balls and F. S. Billett (eds), (1973) *The cell cycle in development and differentiation*, Cambridge University Press.

M. Balls and P. M. Godsell (1973) 'The life cycle of the cell', *Journal of Biological Education*, 7.

R. Baserga (ed), (1971) *The cell cycle and cancer*, Marcel Dekker, New York.

J. D. Ebert and I. M. Sussex (1970) *Interacting systems in development*, Holt, Rinehart & Winston.

R. Fleischmajer and R. E. Billingham (eds), (1968) *Epithelial-mesenchymal interactions*, Williams & Wilkins.

C. F. Graham (1971) 'The design of the mouse blastocyst', *S.E.B. Symposium* 25.

C. F. Graham (1973) 'The cell cycle during mammalian development', in *The cell cycle in development and differentiation*, (M. Balls and F. S. Billett (eds), Cambridge University Press.

A. J. Grove and G. E. Newell (1942) *Animal biology*, 7th edn. (1968), University Tutorial Press.

J. B. Gurdon (1968) 'Transplanted nuclei and cell differentiation', *Scientific American* 219, (6).

J. B. Gurdon (1973) 'Gene expression during cell differentiation', *Oxford Biology Reader* 25, Oxford University Press.

E. Hadorn (1968) 'Transdetermination in cells', *Scientific American*, 219(5).

H. Harris (1970) *Nucleus and cytoplasm*, 2nd edn, Oxford University Press.

D. O. Hall and S. Hawkins (eds), (1973) *Laboratory manual of cell biology*, English Universities Press.

J. R. Hinchliffe (1972) 'Films on animal development', *Journal of Biological Education*, 6.

P. A. Lawrence, F. C. Crick and M. Munro (1972) 'A gradient of positional information in an insect, *Rhodnius*, *Journal of Cell Science*, 11.

D. Malamud (1971) 'Differentiation and the cell cycle', in *The cell cycle and cancer*, (R. Baserga, ed), Marcel Dekker.

C. L. Markert and H. Ursprung (1971) *Developmental genetics* Prentice Hall.

J. M. Mitchison (1971) *The biology of the cell cycle*, Cambridge University Press.

M. A. Monnickendam and M. Balls (1973) 'Amphibian organ culture', *Experientia*, 29.

C. P. Raven (1970) 'The cortical and subcortical cytoplasm of the *Limnaea* egg.' *International Review of Cytology*, 28.

M. R. Redshaw (1972) 'The hormonal control of the amphibian ovary', *American Zoologist*, 12.

M. B. V. Roberts (1971) *Biology: a functional approach*, Nelson.

M. K. Sands (ed), (1970) *The developing organism—a laboratory guide*, Nuffield Advanced Science, Penguin Books.

N. T. Spratt (1971) *Developmental biology*, Wadsworth.

F. C. Steward (1970) 'From cultured cells to whole plants: the induction and control of their growth and morphogenesis', *Proceedings of the Royal Society, series B* 175.

D. Tarin (1972a) 'Tissue interactions in morphogenesis, morphostasis and carcinogenesis', *Journal of Theoretical Biology*, 34.

D. Tarin (ed), (1972b) *Tissue interactions in carcinogenesis*, Academic Press.

J. R. Tata (1973) 'Metamorphosis', *Oxford Biology Reader* 46, Oxford University Press.

L. C. Truesdell and E. C. Cantino (1971) 'The induction and early events of germination in the zoospore of *Blastocladiella emersonii*', *Current Topics in Developmental Biology*, 6.

G. A. Ubbels, J. J. Bezem and C. P. Raven (1969) 'Analysis of follicle cell patterns in dextral and sinistral *Limnaea peregra*', *Journal of Embryology and experimental Morphology* 21.

P. A. Weiss' (1973) 'Differentiation and its three facets: facts, terms and meanings', *Differentiation* 1.

P. A. Weiss, and J. L. Kavenau (1957) 'A model of growth control in mathematical terms', *Journal of General Physiology*, 41.

L. Wolpert (1973) 'Development of pattern and form in animals', *Oxford Biology Reader* 51, Oxford University Press.

H. R. Woodland and C. F. Graham (1969) 'RNA synthesis during early development of the mouse', *Nature*, 221.

Further Reading

C. R. Austin and R. V. Short (eds), 1972. *Reproduction in Mammals*, (vol. 1 'Germ cells and fertilisation'; vol. 2 'Embryonic and fetal development'; vol. 3 'Hormones in reproduction'; vol. 4 'Reproductive patterns'; vol. 5 'Artificial control of reproduction'), Cambridge University Press.

D. R. Garrod (1973) *Cellular Development*, Chapman & Hall.

J. M. Ashworth (1973) *Cell Differentiation*, Chapman & Hall.

D. E. S. Truman (1974) *The Biochemistry of Cell Differentiation*, Blackwells Scientific Publications.

M. Sussman (1973) *Developmental Biology—its Cellular and Molecular Foundations*, Prentice Hall.

Behaviour and neurophysiology

MARIAN DAWKINS

Even the simplest behaviour patterns shown by animals involve many muscles, some acting synchronously and some sequentially. This raises some of the basic questions studied by modern students of behaviour. How do the sense organs, nerves and muscles of the animal interact in such a way that coordinated and functional sequences of movement result? How do these nerves and muscles become 'wired up' during the development of the growing animal so that the right connections are made? At the same time, we might ask how the performance of such a behaviour pattern helps the animal to survive and reproduce: 'Why has natural selection favoured animals which perform this particular behaviour at this particular time?' Looking more closely at how the animal lives in its natural environment, we realise that in order to give a full explanation of its behaviour, we have to take into account that it may live in a complex social group with other animals, that it communicates with its fellows and that its whole way of life is shaped by its environment.

Clearly a subject which takes in such a wide range of complex phenomena is particularly difficult to teach in a way which is both stimulating and comprehensive. The behavioural sciences lack a well-defined theoretical framework and there are few generalisations which hold good for all behaviour. What I shall therefore attempt to do is to pick out, from a wealth of current literature, certain recent discoveries which do seem to reflect important general principles. Perhaps the most important principle in behaviour is still that, like other aspects of life, it has evolved by natural selection. It is as important to bear this in mind when considering why a nerve cell should fire in a particular way as it is when watching a troupe of animals in their natural environment.

I shall also emphasise that the study of behaviour is an

objective and quantitative science; it has tried to free itself from subjective anthropomorphisms about an animal's 'feelings'. The attention which is now paid to experimental design, to objectivity and to the quantitative statistical treatment of data has obvious implications for the design of practical behaviour projects in teaching at all levels.

Sensory input

The problem of how animals respond appropriately to the world about them is not simply the problem of how their sense organs receive stimulation. It is the much more complex and difficult question of how patterns of activity in sensory nerves are processed and transformed into actions. A great tit searching for its food will pick out caterpillars from among twigs and shadows that resemble the food very closely. A bat picks up the returning echo of its own cry and translates it into a three-dimensional picture of objects around it. It is this integration of sensory information which is problematical.

Behavioural investigations

A great deal can be discovered about an animal's sensory capabilities, and the way it processes sensory information, by properly designed and controlled observations of its behaviour towards different sorts of object. A good illustration of this is Hailman's (1971) investigation of why laughing gull chicks peck at the long, thin, red, downward-pointing bills of their parents and not at their parents' legs, which have a very similar appearance and position in their visual field. The chick's pecking stimulates the parent to regurgitate food and the chick can also be induced to peck at models of the parent. Hailman presented thin red needles to the chicks as models of the parent's bill. If he held a needle so that it projected vertically downwards from above and fell only within the upper half of the chick's visual field (as would a real bill), then the chicks pecked at it vigorously. The same needle projecting upwards from below and falling only within the lower visual field was not pecked as much. The most interesting finding was that a needle held so that it crossed the visual field from top to bottom (as the parent's legs might do) was a *less* powerful stimulus for pecking than the needle projecting from above and going only half way. This showed, by purely behavioural means, that there must be a mechanism for integrating information from different parts of the retina, and indeed, these are known physiologically (see below).

Another method is to attempt to train animals to distinguish between two stimuli, rewarding them or punishing them differentially. If they can learn to perform such a task then, provided the experiment has been properly designed, one can conclude that they can distinguish the stimuli, although care must be taken to establish exactly what cues (stimuli) are being used. Failure to learn, however, does not imply that the animals cannot discriminate. This method has been used to establish that bees have colour vision (von Frisch, 1950) and that electric fish (*Gymnarchus*) are able to discriminate minute differences in the ionic concentration of water on the basis of differential distortion of the electric field which the fish creates around itself (Lissman, 1963). It has also been used to show that (in theory) pigeons could use the sun to tell whether they have been displaced to the north or to the south of their home (Whiten, 1972), and that bats can use the delay time between their cry and its echo to discriminate between objects near to them and those only a few centimetres further away (Simmons, 1972).

Neurophysiological investigations
The development of the microelectrode, a tiny wire which can be inserted directly into a nerve fibre and linked to an oscilloscope, has meant that the electrical activity of individual nerve cells can be monitored. Although the following categories are by no means mutually exclusive or comprehensive, they perhaps help to make some sense of the wealth of information on the physiology of sensory systems that has been gained from such microelectrode recordings. Sensory systems may be thought of as adopting one or more strategies (see Barlow, 1961).

1 Specialised, single-purpose mechanisms
Responses are evoked only by specific classes of stimuli ('releasers'). Electrophysiological studies have shown that the olfactory receptors in the antenna of the male silkworm

moth respond almost exclusively to the species-specific sex attractant, bombykol. Single molecules of this substance are sufficient to evoke a response in the antennal nerve (Kaissling and Priesner, 1970), but the moth is apparently able to smell little else, at least not in such minute concentrations. Roeder (1967) has shown that the ears of noctuid moths are specialised to receive the sounds produced by insect-eating bats; the ears are sensitive to the high frequency sound produced by the bat and can distinguish loud from fainter bat cries, but they are little use for anything beyond this one function of bat detection. Lettvin *et al.* (1959) found that cells in the retina of the frog's eye responded specifically to a small round object moving jerkily through the visual field—they seemed to be specialised 'bug-detectors'.

2 *Generalised, multi-purpose mechanisms*

Many animals possess the ability to respond in different ways to many different stimuli. It is unlikely that they have a different specialised pathway from sense organ to brain for each of these responses, since the number of such pathways might have to be prodigious. We therefore find in many animals mechanisms which are used, not just for one purpose, but for many—in catching prey, finding a mate, or for avoiding obstacles during locomotion. The characteristic of such multi-purpose mechanisms is that they respond to features which many different situations will have in common. Barlow (1961) has pointed out that the key to understanding what is important in a wide variety of situations is *change*. Thus a moving object in an otherwise static environment is likely to be of importance because it might denote the change of position of a prey or predator. Many animals, including pigeons and rabbits, have retinal cells which respond only if the animal is looking at an object moving in a particular direction. These cells will not respond to the same object if it is moving in another direction or if it is stationary (Barlow and Hill, 1963; Michael, 1969).

The outlines, or edges of objects, denote change of a different kind—spatial discontinuities between one object and another, or one area of illumination and another. Cells which respond maximally when the animal is shown an edge or a boundary are common in divers animals, including king crabs

(*Limulus*) and rhesus monkeys. Hubel and Wiesel's work on the visual system of the cat (Hubel, 1963) has revealed that the cortex of this animal has a number of such cell types. 'Simple' cortical cells respond if the image of a line falls on the animal's eye, but only if the image is of a particular orientation. Different cells have different preferred orientations, so that some cells respond most to vertical lines, others to horizontal lines, etc. Edges, slits and lines (denoting discontinuities of illumination) are responded to more strongly than solid black or white areas. Another cortical cell type ('lower order hypercomplex' cells) respond only if the animal is looking at a correctly orientated line of a particular length (an operation comparable to that performed by the gull chick discussed above).

The phenomenon of sensory adaptation, whereby the response of a sense organ is initially high when it is first stimulated, but lessens as the stimulation continues, is another example of the importance of '*change of stimulation*' producing the greatest response.

By means of mechanisms which are responsive to changes of stimulation in both time and space, animals have evolved sensory processing mechanisms which can be used in a wide variety of situations and enable them to function efficiently in novel or unpredictable environments.

3 *Changeable-purpose mechanisms* .

Not only are messages carried inwards from the sense organs to the brain, they are also carried outwards from the brain to various points on the sensory pathways. By these messages, the flow of incoming sensory information can be altered before it even gets to the brain, so that the message actually received by the brain may be different on different occasions. In the chick's eye, electrical stimulation of the fibres leading from the brain outwards to the eye makes retinal cells responsive to a wider range of stimuli than when the fibres are unstimulated (Miles, 1970). Microelectrode recordings showed that the responsiveness of auditory nerves (the olivo-cochlear bundle) in the cat, to sounds played to the animal, could be altered by stimulating the fibres running outwards from the brain.

The psychological phenomenon of 'selective attention', for

example as shown in experiments in which human subjects listen to two speech messages, one coming into each ear (Treisman, 1964), could be described as being due to a 'changeable-purpose' mechanism, since the subject usually 'attends' to just one message, largely ignoring the other; but *which* ear is attending can be changed at will. However, this complex switching is probably a central phenomenon (Picton et al., 1971).

These very diverse phenomena all illustrate the basic idea that information received by the sense organs may be processed and utilised very differently at different times, presumably dependent on the current needs of the organism.

Despite the advances which have been made in the investigation of sensory systems, we are still a long way from a full understanding. It is one thing to find an 'edge detector'; it is quite another to be able to explain the complex pattern recognition processes and discriminations which we, and many other animals, perform every moment of our waking lives.

Motor output

What causes the muscles of an animal to contract at the times and in the particular combinations that they do? When we observe an animal perform one behaviour pattern followed by another, the temporal association of these two behaviours could be due to (a) central coordination within the central nervous system, (b) the first behaviour resulting in a change in limb or body posture which then feeds back and acts as the trigger for the second behaviour, (c) patterns of stimulation in the environment which determine when each behaviour pattern occurs; or (d) a combination of these.

Neurophysiologists have now found a considerable number of behaviour patterns which appear to be controlled by preprogrammed 'motor tapes' (Hoyle, 1970), a whole sequence of behaviour being coordinated within the central nervous system. Dorsett *et al.* (1969) showed that the escape response of a sea slug (*Tritonia*), which consists of a series of swimming movements, was centrally patterned. The sequence is controlled by a particular group of cells which show a characteristic pattern of activity when the behaviour is being performed. They show exactly the same pattern when they are almost completely isolated from the rest of the animal, that is, when no feedback from earlier in the sequence, or environmental stimuli, could trigger the later stages. Quite complex behaviour sequences, such as swallowing in dogs (Doty and Bosma, 1956), courtship in grasshoppers (Elsner and Huber, in Hoyle, 1970) and even elements of handwriting in humans (Denier *et al.*, 1965) may also be centrally patterned.

Many sequences show elements of central patterning which is modifiable by various sorts of influences. In the flight control system of the locust, eighty motoneurons drive the muscles of the four wings. A characteristic pattern of nerve impulses occurs in these motor nerves during flight. The same pattern is shown even when the head and these nerves are removed from the rest of the body, and feedback from earlier phases of 'flight' is therefore not available (Wilson, 1968). However, although the basic pattern of nervous activity appears to be centrally determined, it can be modified in various ways: for example, it is speeded up by sensory discharge from nerves in the wings. A similar interaction, between a central programme and various sources of input, has been found in many instances (Hoyle, 1970; Evoy and Cohen, 1971).

External influences, such as the behaviour of another animal, are obvious reasons why an animal should change from one behaviour to another. Tinbergen's (1951) study of stickleback courtship, where male courtship activities elicit female responses, which in turn evoke the next male activity, is a classic example. Reese (1963) studied the stimuli which elicited the sequence of movements shown by hermit crabs selecting a new shell, and argued that their behaviour can be understood as a series of responses to various aspects of the shell as they are sequentially encountered.

Although we may reasonably hope soon to have a full explanation of relatively simple behaviour patterns, to understand, for example, exactly how the eighty moto-neurons controlling locust flight interact to produce their effect, it will obviously be a long time before we have a comparable understanding of more complex behaviours in animals with more sophisticated nervous systems. Indeed, trying to give an 'explanation' of behaviour in terms of the activity in individual nerves for a vertebrate with a total of about 10^{10} neurons may be not just impossible, but undesirable. 'Higher level' explanations have therefore often been evoked to explain the behaviour of complex animals, including concepts such as 'motivation' and 'drive'. Sexual motivation or drive might be used to describe the (as yet unknown) factors which seemed to be responsible for the appearance of several different components of courtship. It is possible to be rigorous in the use of such terms, by use of statistical techniques such as factor analysis (Wiepkema, 1961), to decide whether different behaviours are closely correlated with each other so that it is reasonable to assume they have many causal factors in common.

Although somewhat unfashionable of late, motivational models have recently been given a more concrete and quantitative basis by McFarland (1971) who has successfully applied engineering concepts to behaviour. In particular, his use of control theory, and his emphasis on the fruitful comparisons to be made between the physical and biological sciences, may give the explanation of complex behaviours the theoretical framework it so badly needs.

Genetics of behaviour
As behaviour involves the coordinated control of a number of different parts of the body, it is not surprising to find that a given behavioural trait is affected by many different genes. Variations in behaviour due to single genes are comparatively rare (but see Bastock, 1956), and quantitative methods for studying polygenic variation between animals are usually necessary (Fuller and Thompson, 1960; Parsons, 1967). The commonest techniques are: using inbred strains maintained by brother-sister mating over many generations; and using artificial selection. These techniques have now been applied to many different species.

More recently developed techniques hold out the promise of being able to pin-point where a given mutation is exerting its primary effect in the organism (Hoffa and Benzer, 1972). By making use of aberrant 'chimaera' individuals, which have a different genetic constitution in different tissues of their bodies, the effect of different mutations on particular areas can be isolated (Ikeda and Kaplan, 1970).

Development of behaviour

The development of any trait, behavioural or otherwise, involves a complex interaction between genetic and environmental determinants. Nerves, muscles and sense organs must develop, and somehow connect with each other, to produce the functioning behaving organism. It has recently been found that particular types of sensory experience in young animals may have a profound effect upon how their nervous systems develop. The fact that cats have cortical cells specifically sensitive to lines of particular orientations has already been mentioned (p. 80). If young kittens are reared so that they only *see* lines of one orientation, they develop cells sensitive to this one orientation, whereas a normal kitten will have some cells sensitive to each of several different orientations (Blakemore and Cooper, 1970). An exposure of just one hour to lines of particular orientation is sufficient to alter the constitution of the cortex (Blakemore and Mitchell, 1973), provided the exposure is done at a 'critical period' in the animal's life.

The idea of a critical period, when the young animal is held to be particularly vulnerable to certain stimuli, is also found in Lorenz's classical ethological studies on imprinting. Here, exposure to a parent-like stimulus in early life was held to affect, not only the likelihood that the young bird or mammal would follow this object when juvenile, but also its choice of sexual partner in later life. The lasting effects of early experiences on subsequent social behaviour have been shown in many animals, including ducks, rats and rhesus monkeys (Harlow and Harlow, 1962).

Particular sorts of experience also affect the development of co-ordinated movements in an animal. Kittens which have plenty of visual experience of the world, but are physically confined so that they have no chance to associate changes in what they see with the results of their own movement, are very abnormal. Both their blinking responses and their ability to reach out and touch objects with their paws are upset, not through any failure of their eyes or limbs, but because they have never been able to coordinate the two (Held and Hein, 1963).

A word should perhaps be said about the description of a behaviour pattern as 'innate' as it is still controversial among ethologists whether it is useful (or possible) to distinguish innate from learned elements of behaviour. Clearly, both genetic and environmental influences affect all behaviour, and in this sense, no behaviour is entirely genetically determined. However, some ethologists still like to use the term 'innate' to cover cases where behaviour develops in the absence of particular sorts of experience (Lorenz, 1965). For example, young pied flycatchers show the species-typical owl-mobbing response even when they have never seen an owl before (Curio, 1970). To describe this as 'innate' would mean, not that the environment had been without influence on their development, but simply that the particular environmental influence of 'previous experience with owl' was not necessary for the mobbing behaviour to be shown.

Social organisation and communication
The study of animal societies, how they are structured and how the animals within them communicate with each other, has been greatly helped by the development of technical aids, such as radio-tracking, and of statistical techniques for dealing with observations that are collected. It is no longer sufficient to provide anecdotes of particular incidents. Quantitative measurements are needed to reveal the determinants of social behaviour. These may involve such things as the numbers of times animals are seen in association with each other, their spatial distribution and their group size. For example, it has been found that spotted hyaenas, living in an area where there are many zebra, exist in larger packs than those living where the prey is smaller and can be scavenged or caught by a smaller group of animals (Kruuk, 1972). By collecting quantitative data, it was possible to assess the functional significance of different pack sizes to food supply. Similarly, differences in social behaviour in different species of animal may often be correlated with measurable differences between some aspect of their environment, such as type or distribution of food or predation. Such correlations have been made in birds (Lack, 1968), carnivores and primates (Crook and Gartlan, 1966; Jolly, 1972), amongst others, illustrating the complex interactions existing within ecological communities.

The signal value of a movement may often be initially

established by statistical correlation of the behaviour of one animal with that of another. Hazlett and Bossert (1965) were able to show that certain behaviours (such as leg-raising) in one hermit crab were reliably followed by statistically significant increases or decreases in behaviours shown by a second crab. In this way they could demonstrate objectively, and without being anthropomorphic, that the crabs were signalling to each other. The subject of animal communication is well reviewed by Cullen (1972).

Teaching animal behaviour
Most British universities now offer animal behaviour and neurophysiology as part of their Zoology courses, although both the emphasis and the degree of compulsion varies (see *Which University*). In addition, other degree courses, such as psychology (e.g. Bristol), psychology-zoology (e.g. Durham and Manchester) and the human sciences course in Oxford include animal behaviour as an important section.

Universities also vary considerably in the way in which they present and integrate behaviour and neurophysiology. I am most familiar with the course at Oxford, where a primary aim of behaviour teaching is to emphasise that in order to survive and reproduce, animals have to overcome certain 'problems': the problem of recognising patterns of stimulation in the environment, the problem of communicating over a long distance, and so on. This approach is applicable at the neurophysiological level as well as at the overt behavioural level and, it is hoped, helps to unify various approaches to the study of behaviour. For practical work, undergraduates are encouraged to work on their own research projects and to discover some of the problems involved in behaviour research, of which there are many. The fact that many animals behave differently at different times of the day, or depending on whether they are alone or with others, means that considerable thought must be given to the design of experiments and the conclusions that can be drawn from them. There may be individual differences between animals of the same species or variations in the behaviour of a single individual in apparently identical conditions on two occasions. The high degree of variability in behaviour means that considerable reliance has to be placed on statistical analyses (e.g. Siegel, 1956), and the use of these is encouraged.

Not so very long ago, behaviour appeared in school biology only under the somewhat curious heading of 'Irritability'. A major trend in the teaching of biology has been to present behaviour not as an extra, tagged on to an animal as an afterthought, but as an integral part of its equipment for survival. The Nuffield O-level course, for example, includes sections on behaviour as such, and also discusses anatomy and physiology in their dynamic roles as the machinery of behaviour. Nevertheless, relatively little animal behaviour is taught in schools.

In considering how much animal behaviour should be included in the school curriculum, it is important to ask what behaviour studies can contribute to a biological education and indeed to education generally.

It is often argued that animal behaviour is important because it can teach us something about ourselves. Can we not look to other animals for the 'roots' of our own behaviour? If we want to understand human aggression, for example, should we perhaps study aggression in other species, just as we assess the likely effects of drugs on the human body by their effects on rodents? Indeed, a number of popular books on animal behaviour have made just such extrapolations. Unfortunately, however, we do not have the behavioural equivalent of the guinea pig. This is partly because our understanding of the behaviour of any one species is so fragmentary and partly because behaviour, perhaps more than biochemistry or physiology, seems so variable from species to species. Ethologists are rightly chary of generalising from the behaviour of the ten-spined to that of the three-spined stickleback. There are major differences in aggressive, courtship and other behaviours between the two species, many of them connected with the fact that the spines of the three-spined species are a more effective defence against predation (Hoogland *et al.*, 1956; Wilz, 1971). Similarly, major behavioural differences exist among our closest relatives, the apes. Chimpanzees, gorillas, gibbons and orangs all have very different social behaviours and ways of life, and may have evolved so differently from ourselves that none of them could be taken as a reliable model from which to extrapolate our own behaviour.

This is not to say that there are no parallels to be drawn between ourselves and other species. Animal studies provide hypotheses to explain our own behaviour even if we recognise fundamental differences. Recognition of similarities between animal and human signals, for example, can give a new perspective to a familiar phenomenon. Like other species, ours has been shaped by natural selection. Behaviour, morphology, and physiology reflect adaptation of organisms to environment.

The practical study of behaviour presents certain problems mentioned earlier, but these can also be looked upon as opportunities. The need in such work for objectivity, for correctly designed experiments and for statistical treatment of data can be used to explain the rationale and methodology of science in general as well as being specifically instructive in behaviour.

Behaviour could also theoretically provide valuable material for communicating ideas on evolution and natural selection. However, there are certain dangers in this if fundamental concepts come to be misunderstood rather than clarified. For example, an important topic in evolutionary theory is whether animals are 'altruistic' or 'selfish'. If behaviour is the result of natural selection, then animals will behave in such a way as to ensure their own individual survival, or that of their offspring (Darwin, 1859) or their close relations (Hamilton, 1964). The statement 'In higher animals, behaviour may take the form of individual suicide in order to ensure the survival of the species' (Nuffield Biology Teachers' Guide IV) is a serious misunderstanding of the theory of natural selection. G. C. Williams (1966) has provided an excellent and readable coverage of the way in which natural selection operates. Provided that they are handled knowledgeably, discussions of the role of behaviour in evolution can be very valuable.

I have rather stressed the disadvantages of teaching behaviour in schools as I think there are genuine difficulties involved in the description, explanation and interpretation of behaviour. To teach ethology without due regard for these difficulties, for example, to make oversimplistic extrapolations from the behaviour of other species to that of our own (or vice-versa), may in fact do more harm than good. However, I hope I have also made it clear that the disadvantages can, in skilful hands, be turned into advantages. Familiarity, interest and ease of observation all commend behaviour as a school subject. A biological education which has stressed molecules, cells and sense organs without showing what animals *do* with these component parts is clearly incomplete. Similarly, the general education of biologist and non-biologist alike can be greatly enriched by using behaviour studies to introduce modern concepts of evolution and natural selection, and to explain the methodology of science in general.

References

H. B. Barlow (1961) 'Possible principles underlying the transformation of sensory messages', in *Sensory Communication*, W. A. Rosenblith (ed.), M.I.T. Press.

H. B. Barlow and R. M. Hill (1963) 'Selective sensitivity to direction of movement in ganglion cells of the rabbit retina', *Science*, 139.

M. Bastock (1956) 'A gene mutation which changes a behaviour pattern', *Evolution*, 3.

C. Blakemore and G. F. Cooper (1970) 'Development of the brain depends on the visual environment', *Nature*, 228.

C. Blakemore and D. E. Mitchell (1973) 'Environmental modification of the visual cortex and the neural basis of learning and memory', *Nature*, 241.

J. H. Crook and J. S. Gartlan (1966) 'On the evolution of primate societies', *Nature*, 210.

J. M. Cullen (1972) 'Some principles of animal communication', in *Non-verbal Communication*, R. A. Hinde (ed.), Cambridge University Press.

E. Curio (1970) 'Kaspar-Hauser-Versuche zum Feinderkennen junger Trauerschnäppner (*Ficedula h. hypoleuca* Pall.)', *Journal für Ornithologies* 111.

C. Darwin (1859) *The Origin of Species*, Oxford University Press.

J. J. van der Gon Denier, and Ph. J. Thuring (1965) 'The guiding of human writing movements, *Kybernetik*, 2.

D. A. Dorsett, A. O. D. Willows and G. Hoyle (1969) 'Centrally generated nerve impulse sequences determining swimming behaviour in *Tritonia'*, *Nature*, 224.

R. W. Doty and J. F. Bosma (1956) 'An electromyographic analysis of reflex deglutition', *Journal of Neurophysiology*, 19.

W. H. Evoy and M. J. Cohen (1971) 'Central and peripheral control of Arthropod movements', *Advances in comparative Physiology and Biochemistry*, 4.

K. von Frisch (1950) *Bees, their vision, chemical senses and languages*, Ithaca.

J. L. Fuller and W. R. Thompson (1960) *Behaviour Genetics*, John Wiley.

J. P. Hailman (1971) 'The role of stimulus-orientation in eliciting the begging response from newly-hatched chicks of the laughing gull (*Larus atricilla*)', *Animal Behaviour*, 19.

W. D. Hamilton (1963) 'The evolution of altruistic behaviour', *American Naturalist*, 97.

H. F. Harlow and M. K. Harlow (1962) 'Social deprivation in monkeys', *Scientific American*, 207.

B. A. Hazlett and W. H. Bossert (1965) 'A statistical analysis of the aggressive communications system of some hermit crabs', *Animal Behaviour*, 13.

R. Held and A. Heir (1963) 'Movement-produced stimulation in the development of visually guided behaviour', *Journal of comparative and physiological psychology*, 56.

R. Hoogland, D. Morris and N. Tinbergen (1956/57) 'The spines of sticklebacks (*Gasterosteus* and *Pygosteus*) as means of defence against predators (*Perca* and *Esox*)', *Behaviour*, 10.

Y. Hotta and S. Benzer (1972) 'Mapping of behaviour in *Drosophila* mosaics', *Nature*, 240.

G. Hoyle (1970) 'Cellular mechanisms underlying behavior — neuroethology', *Advances in Insect Physiology*, 7.

D. H. Hubel (1963) 'The visual cortex of the brain', *Scientific American*, 209 (5).

K. Ikeda and W. D. Kaplan (1970) 'Unilaterally patterned neural activity of gynandromorphs, mosaic for a neurological mutant of *Drosophila melanogaster*, *Proceedings of the Nat Society*, 67.

A. Jolly (1972) *'The Evolution of Primate Behaviour'*, Macmillan, New York.

K. E. Kaissling and E. Priesner (1970) 'Die Riechschwelle des Seidenspinners', *Die Naturwissenschaften*, 57.

H. Kruuk (1972) *The Spotted Hyena*, University of Chicago Press.

D. Lack (1968) *Ecological Adaptations for Breeding in Birds*, Methuen.

J. Y. Lettrin, H. R. Mantura, W. S. McCulloch and W. H. Pitts, 'What the frog's eye tells the frog's brain', *Proceedings of the Institute of Radio Engineers*, 47.

H. W. Lissman (1963) 'Electric location by fishes', *Scientific American*, 208 (3).

K. Lorenz (1965) *Evolution and Modification of Behavior*, University of Chicago Press.

D. J. McFarland (1971) *Feedback Mechanisms in Animal Behaviour*, Academic Press.

C. R. Michael (1969) 'Retinal processing of visual images', *Scientific American*, 220 (5).

F. A. Miles (1970) 'Centrifugal effects in the avian retina', *Science*, 170.

P. A. Parsons '1967) *The Genetic Analysis of Behaviour*, Methuen.

T. W. Picton, S. A. Hillyard, R. Galambos and M. Schiff (1971) 'Human auditory attention: a central or a peripheral process?', *Science*, 173.

E. S. Reese (1963) 'The behavioural mechanisms underlying shell selection by hermit crabs', *Behaviour*, 21.

K. D. Roedei (1967) *Nerve Cells and Insect Behaviour*, Harvard University Press.

S. Siegel (1956) *Nonparametric Statistics for the Behavioral Sciences*, McGraw-Hill.

J. A. Simmons (1971) 'Echolocation in bats: signal processing of echoes for target range', *Science*, 171.

N. Tinbergen (1951) *The Study of Instinct*, Clarendon Press.

A. Treisman (1964) 'Selective attention in Man', *British Medical Bulletin*, 20.

'Which University' (1972) Cornmarket Press.

A. Whiten (1972) 'Operant study of sun altitude and pigeon navigation', *Nature*, 237.

P. R. Wiepkema (1961) 'An ethological analysis of the reproductive behaviour of the bitterling', Arch. neerl. Zool. 14.

G. C. Williams (1966) *Adaptation and Natural selection, a critique of some current evolutionary thought*, Princeton.

D. M. Wilson (1968) 'The flight-control system of the locust', *Scientific American*, 218 (5).

K. J. Wilz (1971) 'Comparative aspects of courtship behavior in the Ten-spined stickleback', Z. Tierpsychol, 29.

Further Reading

V. B. Droschei (1969) *The magic of the senses: New discoveries in animal perception*, W. H. Allen.

A. Manning (1972) *An introduction to animal behaviour*, 2nd edition, Edward Arnold.

N. Tinbergen (1965) *Social behaviour in animals with special reference to vertebrates*. Methuen, and Science Paperbacks.

N. Tinbergen (1966) *Animal Behaviour*. Time-Life International (Nederland).

Population genetics and evolution

DAVID TOMLEY

Organisms seldom live in isolation. They generally occur in groups in which free and random interbreeding occurs. Such a group is an *interbreeding population* or *population* and theoretically every member of it is just as likely to breed with one member as with another, but has very much less chance of breeding with a member of another group. For example, rats living in a granary on an isolated farm will have only a remote chance of mating with rats from elsewhere. Similarly grain weevils in a particular sack of wheat are far more likely to interbreed with weevils from the same sack. If plants are considered then cross pollination is most likely to occur between plants in the same population as the bees work the flowers in that area.

While the individual has a limited lifespan and a fixed set of genes or genotype, a population is virtually immortal and its total genetic make-up may gradually or suddenly change. This process of cumulative genetic change within a population is the process of *evolution*. It is the population which is the unit of evolution, and it is the genetics of the population that deserves the close attention which traditionally has been given to the inheritance patterns of the individual.

A chapter as brief as this can only hope to touch on some of the areas that have resulted in the present emphasis on populations and the mechanism of evolution. The bibliography and references at the end of the chapter provide a more comprehensive survey of the subject and examples of the investigations responsible for our present knowledge.

The gene-pool and sources of variation
The *gene-pool* comprises the total number of genes available in a population at any one time. It indicates not only the types of genes present but also their frequency. It is from this supply of genes, in the form of gametes, that the

next generation is selected. All natural populations are genetically variable, each member has a particular genotype which is unlikely to be the same as that of any other member. It is this genetic variability which enables the population to change should the prevailing conditions alter.

In sexually reproducing organisms *meiosis* results in expression of this variation due to the independent assortment of chromosomes and the recombination of linked genes, where crossing-over occurs. *Cross-fertilisation* will result in new combinations of genes appearing in every generation and it is upon this heritable variation that natural selection acts. New sources of variation, in the form of mutated genes, are slowly but continually added to the gene-pool. A *mutation* occurs in an individual but it can have no evolutionary significance until it has entered the gene-pool. If the mutation is a recessive one then it will not be expressed in the phenotype until two such mutants occur together upon pairing of homologous chromosomes at fertilisation. For this to occur the original mutant must have spread through much of the population by descent from the original mutation, though the rate of spread may be increased by the same mutation occurring spontaneously in other individuals. As the numbers of mutant homozygotes in the population increases they will be selected *for* if the gene combination is more advantageous than the existing one under present conditions. They will be selected *against* if the homozygous condition is less favourable to survival. In any case a mutation has only a fifty-fifty chance of being beneficial; therefore, in populations already well adapted to their environments, any mutation is likely to be disadvantageous.

The variability within a population is also affected by the type of mating system. An *inbreeding* system of self-fertilising organisms, or a laboratory population resulting from a long series of *sib* matings (brother and sister), will be largely homozygous for all genes. This means that little or no variability is present for natural selection to act upon, so evolutionary changes will be at a standstill. The tendency to homozygosity in inbreeding systems can be simulated by tossing a coin. Tossing the coin twice gives a genotype for one pair of alleles. For example, let heads represent gene A and tails its allele a. Then heads-heads sequence gives AA, heads-tails or tails-heads Aa, and tails-tails aa. The probability of throwing a head is ½. With the second throw the probability is again ½. The chance of throwing one head followed immediately by another is therefore ½ x ½ = ¼ since these two events are unrelated. So the probability of throwing

$$\begin{aligned} AA &= ¼ \\ Aa &= ¼ \\ aA &= ¼ \\ aa &= ¼ \end{aligned}\left.\begin{aligned} \\ \\ \end{aligned}\right\} ½$$

AA and aa can give rise only to progeny like themselves, that is, to more homozygotes. The offspring of heterozygotes Aa and aA will be half homozygotes at every generation. This leads to a rapid diminishing of the numbers of heterozygotes with time. We have considered only one locus but what is true for one is true for them all.

One way in which variation is introduced into an inbreeding system is through mutation, but this low rate of genetic change is sufficient to bring about evolutionary change only if the mutation occurs in a genotype where its effect is advantageous.

Compared with an inbreeding system where the amount of variation tends to be reduced, an *outbreeding* system of cross-breeding organisms ensures genetic variablity, not only by maintaining the heterozygotes and any mutated genes that occur, but also by combining this variability in new genotypes as a result of meiosis and fertilisation.

Gene frequency

When the conditions under which a population exists remain more or less stable, and this is the usual short-term state of affairs, it is likely that the genes and their relative proportions in the gene-pool will remain the same from generation to generation. The number of a particular allele present in the gene-pool, expressed as a percentage of the total number of available loci in the population, is known as the *gene frequency*. It can be represented as follows:

From a pair of alleles, A and a, the possible genotypes in a population will be AA, Aa, aa. Then, since this represents all the genes, A and a, in the population p + q = 100% or 1

(unity), where p = frequency of gene A
and q = frequency of gene a

Every egg and every sperm in the gene-pool will have either
allele A or a, so the genotypes that can form will be

sperm \ egg	A	a
A	A A	Aa
a	Aa	a a

and their frequencies can be found by substituting p and q

	p	q
p	p^2	pq
q	pq	q^2

so the possible genotypes are AA, Aa, aa, occurring in the proportions of p^2 :2pq:q^2 respectively.

These genotypes in these frequencies provide the gene-pool for the succeeding generation. The genes are the same, A and a, and their new frequencies can be calculated as follows:

let the numbers of the genotype   AA be C

that of   Aa be D

and that of   aa be E

then the frequency of allele A in the population is

$$\frac{2C + D}{2\,(C + D + E)} \quad \text{or} \quad \frac{C + \tfrac{1}{2}D}{C + D + E}$$

since homozygotes (AA) carry two of allele A, the heterozygotes (Aa) carry one A, and the total number of available loci is twice the number of individuals in the population (since each has two loci for this gene).

That the gene frequency tends to remain the same under certain conditions was first shown to be true by G. H. Hardy and G. Weinberg, working independently in 1908. They considered a large, randomly interbreeding population—a *panmictic* population—in which there were:

no mutation occurring;

no movement of members of the population into or out of the area (no immigration or emigration);

no selection, with all genotypes equally likely to survive.

Considering one pair of alleles, A and a, where the frequency of gene A = p and of a = q:

$$\text{frequency of A gametes/genes} = p^2 + \tfrac{1}{2}\,(2pq)$$
$$= p(p + q)$$
$$\text{frequency of a gametes/genes} = q^2 + \tfrac{1}{2}\,(2pq)$$
$$= q(p + q)$$

Dividing both by (p + q), this leaves the frequency of gametes bearing gene A (or the frequency of gene A) as p, and the frequency of a as q, which is the same as before.

The proportion of the two alleles in the genotypes of ensuing generations will not change; so that in a large panmictic population with no selection, the expectation is that the initial gene frequency will remain unaltered. When these conditions prevail, the population is said to be in *Hardy-Weinberg equilibrium*.

With changing environmental conditions, however,

different phenotypes and therefore genotypes may be selected. The sum total of genes from these new genotypes will form the new gene-pool in which some gene frequencies may be significantly different from those in the parental gene-pool. For example, the recessive gene Hb.S. (haemoglobin S), which results in the incorporation of the amino-acid *valine* instead of *glutamic acid* at the sixth position in the B-chains of the haemoglobin molecule, confers a selective advantage on people carrying it in that they are markedly more resistant to the malarial parasite. This gene is known to occur in up to 20 per cent of much of the population of West Africa. In the present negro population of the southern United States of America which is derived from the African parental population, but lives where malaria does not occur, this gene is found in only 9 per cent of the population. Thus an evolutionary change in the population occurs, resulting from a change in the frequency of the genes.

Selection

In 1858, both Darwin and Wallace suggested that the main controlling factor producing evolutionary change was *natural selection*. Their argument was essentially that organisms tend to produce more offspring than the environment will support; that the numbers of organisms tend to remain constant, therefore many must die before they reproduce. Those that survive will include among them the phenotypes best suited to the existing environment. This means that over a period of time optimum phenotypes will become commoner in the population.

These arguments are the same today; overproduction of genetically variable offspring, leading to competition for limited resources in the environment, resulting in the selection of the best-suited phenotypes or the elimination of those members least well-adapted to their particular environment. For example, the adaptive immunity of the warfarin-resistant strain of brown rat found on the Shropshire/Montgomeryshire border, and of the insecticide-resistant strains of housefly, mosquito, bedbug and cockroach found in many areas, has meant that these organisms survive because they are better adapted.

Some of the controlling factors operating in natural

selection can be listed readily: density of organisms; climatic or micro-climatic factors, such as light, water and temperature requirements, may be unsuitable; there may be insufficient nesting sites; not enough food either in the form of prey, desired food plant, or minerals in the soil; alterations in the physiology or biochemistry of an organism, resulting perhaps in sterility, or lack of an enzyme or enzyme system, causing loss of an essential pigment, such as chlorophyll or melanin; changes resulting in rapid growth, a bigger organism, or a more effective behaviour pattern; changes in cryptic colouration; or greater intelligence. All these may ensure that such phenotypes are included among those that survive. In spite of such a list, the factors that control and regulate populations are incompletely understood. What is clear, however, is that the organisms in a population are subject to selection pressures which are determined by the controlling factors. As the controlling factors vary, so do the selection pressures. In a given situation under a certain set of selection pressures, particular combinations of genes have a loading or selective value. This value will vary with the pressures operating. It used to be thought that selection pressures were gentle, but it is now known that this is not so. A change in selection pressures can account for a 40 per cent change in a population fairly easily and values up to 80 per cent have been recorded on occasions. (Dowdeswell, 1971.)

It is fair to say that if a certain variety is common in a population then it is well adapted to the existing environment. If a new variety becomes commoner at the expense of the other then it may be better suited to the old environment. Alternatively there may have been a subtle change in the balance of controlling factors which makes the new variety better suited to the new environment. Many suggestions may be made about why one variety is better suited to its environment than another, but in the end suitability or fitness is measured by the numbers of that variety in the population, that is, by its reproductive fitness.

Types of selection
Selection acts on a genetically variable population, producing one which is adapted to the existing environment. The short-term pattern is one in which selection maintains the

stability of a well-adapted population within a reasonably constant environment. This *stabilising selection* is vital and fundamental. Too little variation will mean that if conditions change rapidly there may be no members of the population that survive. Too much variation will mean that few of the existing population are well adapted to present conditions and, if these conditions continue, the population may rapidly decrease in size or even die out. When a new gene combination, or a single dominant gene mutation, proves to have a high loading then the previously common type will be replaced by the new. The former gene frequencies change with resulting *directional selection* (see fig. 1).

Selection pressures may favour the extremes of variation rather than the mean; so that intermediates are selected against. What was a single population, with a wide range of variation, becomes separated into two populations, each with more limited variation. This *disruptive selection* may result in the formation of new sub-species, or even species.

Selection in action
Selection operates on the available genetic variation in the population to produce a gene-pool with genes more suited to the environment than were the genes of the gene-pool of the preceding generation.

It was mentioned earlier that natural selection is responsible for the stability of populations as well as for the changes that occur within them. The best examples of stabilising selection are possibly to be found in environments which change very little with time. One of these is the sea. The survival of *Latimeria* sp., the coelocanth, from the Devonian period to the present day, indicates the selective advantages of its particular gene combinations in an environment that has shown little change over millions of years. Perhaps the Loch Ness monster is a similar example.

1 THE PEPPERED MOTH
The classic example of *directional selection* is that of the spread of the gene which gives rise to the *carbonaria* form of *Biston betularia*, the peppered moth. This example is illustrative of a dominant mutation becoming advantageous with a change in selection pressures. The influence of man

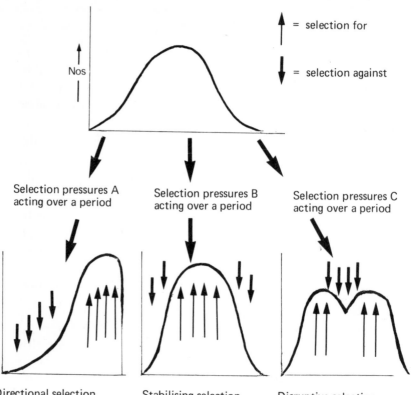

= selection for

= selection against

Selection pressures A acting over a period

Selection pressures B acting over a period

Selection pressures C acting over a period

Directional selection.

New variety at selective advantage due to higher loading in existing environment, or previously uncommon, or new variety at selective disadvantage in changed or changing environment.

This type of selection results in a new form replacing the old.

Stabilising selection.

Extreme variations in either direction from the norm are selected against. Gene combinations well-adapted to existing conditions are at selective advantage.

This type of selection maintains the *status quo*.

Disruptive selection.

Either extreme is equally suited to the new condition. The intermediates have a low loading and are at a selective disadvantage.

This type of selection may result in the formation of two sub-species and eventually two species.

Figure 1

can be of paramount importance in altering the environment and so affecting selection pressures. This happened in the case of the melanic variety of the peppered moth. The gene for this form spread rapidly through populations of the moth in industrial Britain, where its dark colour offered better protection from predators. (See Sheppard, 1967.) Evidence of the nature of controlling factors (or selective agents) is very difficult to acquire since the genes under consideration may have multiple effects. It is quite possible that the visible effect is less important than some secondary physiological or biochemical effect. A second possibility is that gene A (or gene series A if polygenes are being studied) is closely linked with gene B and that gene A has a visible, but unimportant effect, whilst gene B has a non-visible effect of some magnitude and that it is this which is important.

Kettlewell's superb proof (see Kettlewell, 1958) of predation by birds may well be exceptional. It also appears that in some species of moth the genes, giving rise to melanism in the adult, confer upon the larva the ability to survive in conditions of more limited food supply than those tolerated by non-melanic forms. This is an important consideration, for not only does it mean, in this specific case, that the melanic form should spread in non-industrial areas, provided that the larvae advantage is not overriden by the adult disadvantage, but also that genes may have secondary effects, not always as obvious, but often more important than the first observed effect. In addition, this shows that selection in the young form may be in one direction while selection in the adult may be in another. This is known as *endocyclic selection*. The spread of the gene causing melanism in the peppered moth is also an example of how rapidly the gene frequencies in gene-pools can change. Its rate of increase since the industrial revolution must have meant a selective advantage for the *carbonaria* genotype of the order of 30 per cent each year. In this case the reason for the selective advantage of the gene is clearly due to man's activities and the consequent alteration of the environment. When more than one form of the same species occur together in a population, and the rate of mutation alone is insufficient to account for the numbers of the rarest form, then *polymorphism* exists. When one gene is replacing another in a population, and the genes give rise to different forms, then a *transient polymorphism* will occur during this time. A *balanced polymorphism* occurs when opposing sets of selection pressures act to maintain different forms in the population. Pin-eyed and thrum-eyed primroses are good examples of this, as are background colour and banding patterns in the snail *Cepaea,* human ABO blood groups, sickling in red blood cells, and cyanogenesis in birdsfoot trefoil, *Lotus corniculatus*, and white clover, *Trifolium repens*. (Dowdeswell, 1963; Nuffield Advanced Science, 1971; Ford, 1964.)

2 *THE MEADOW BROWN BUTTERFLY*

If selection pressures are eased a population tends to increase in numbers, with a consequent increase in variation among its members. Decrease in numbers in a population is indicative of increased selection pressures, with a consequent decrease in variation. Selection can often be seen in operation, where the environment is undergoing change, by studying populations at the extremes of their geographical range. Such a study has been continuing for over twenty years in the Meadow Brown butterfly, *Maniola jurtina,* carried out by Ford, Fisher, Dowdeswell, Creed and McWhirter. Summaries of some of the work may be found in both Dowdeswell (1963) and Ford (1964, 1965).

A few points have been selected for discussion below.

The underside of the hind wing of *Maniola jurtina* bears a series of spots, numbering from 0 to 5. The spot distribution pattern in most of southern England, the southern English stabilisation, is that the males are unimodal at 2 spots and the females unimodal at 0. This pattern is part of a larger European stabilisation, extending from Scandinavia to the Black Sea, and across to Portugal. As stated earlier, at the edge of its range a species tends to show more variation, and this applies to the spot pattern in *Maniola*, which changes in an area near Launceston (on the Devon-Cornwall border) before once more settling down into a different pattern, the East Cornish pattern, with the males still unimodal at 2 spots, but with females bimodal at 0 and 2 spots, the greater mode being at 0 spots. This change is an abrupt one, sometimes occurring over a distance of a few yards. It involves no physical barrier and the site of the change, from the southern

English to the East Cornish pattern, has moved eastwards, then westwards, over a distance of some forty miles in the past thirteen years. The two populations remain distinct despite the inevitable mixing of genes, the *gene-flow*, that must occur between the two populations. This is an example of *disruptive selection*, and sympatric evolution, though the controlling factors are not clear. Since 75 per cent of the life span of the Meadow Brown butterfly is spent as a caterpillar, it is likely that the selection operates at that stage (Dowdeswell 1971). Work by McWhirter and Scali is investigating the relationship between intestinal bacteria in the caterpillar and adult spotting, though it may be that the spotting pattern is related to the susceptibility of the larva to parasitisation by the Braconid *Apanteles tetricus*.

Ford (1965) says 'The subdivision of a species into small isolated units also favours its rapid evolution. This it does because each group can then fit itself to the special ecology of its own locality whereas a population occupying a large continuous area can be adjusted only to the average of the conditions which obtain there.' Once again, work on the Meadow Brown butterfly illustrates this point. In the Scilly Isles, populations of this organism have been studied on five small islands, of less than forty acres, and on three larger islands, each of more than 680 acres. While the populations on all three larger island shared the *same* spotting pattern, those on the five small islands differed among themselves over the same period of time. This would seem to suggest that populations of the same species occupying large and diversified areas are usually more alike than populations occupying separate small areas. In the small areas, each population will be adapted to its own local conditions and it will interact more closely with the controlling factors. This sort of explanation can also be applied to Darwin's *Finches of the Galapagos Archipelago*.

A variety of sources suggest that small ecological barriers, like a ploughed field, a high hedge, a wood, or a field of short grass, can act as agents which effectively isolate populations. This has also been demonstrated effectively in the case of the Meadow Brown butterfly.

In 1961 one of the Scilly Isles, White Island, was cut into

two by the sea. A single population of *Maniola* became two isolated ones and the spotting pattern on the hind wing of the butterflies is changing. An even more spectacular example of the speed with which selection acts, and therefore of the high selection pressures, was shown by the female butterflies on Tean Island. These had a bimodal spot distribution, with 2 as the greater mode and 0 as the smaller. In one season this changed to an entirely unimodal distribution at two spots (see fig. 2). This change was associated with the removal of a herd of cattle from the island. The effect on the grass was to change it from a closely cropped sward to a hayfield, so that a considerable alteration in selection pressures ensued; a different gene combination had a more favourable loading in the changed environment, and was therefore at a selective advantage. The alteration of gene frequencies in the gene-pool took less than a year to bring about this change.

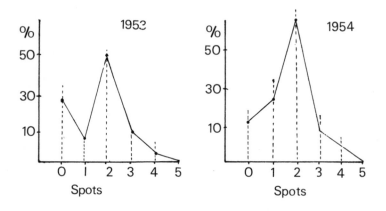

After Ford, *Mendelism and Evolution,* Methuen 1965

Figure 2

Trends in University teaching
Emphasis in university courses on population genetics is now increasing rapidly. With integrated schools of biological sciences rather than separate departments of botany and zoology, courses drawing on both plants and animals as illustrative material are more common.

Many university courses are sub-divided into blocks, units or modules from which the students themselves build their own course and it is not unusual to find basic courses in population genetics and also more advanced ones.

There has also been closer integration of theoretical and practical aspects of the work so that students gain insight into the sorts of investigations which have led to the present state of our understanding of evolution.

Seminars, practical classes and lectures play their part in the courses which vary in length from six weeks to half a year. The courses invariably include the Hardy-Weinberg equilibrium, gene frequencies and the factors influencing these, selection, polymorphism, isolation and speciation.

Some courses use computer simulations of population studies. The aim of the courses is to illustrate the mechanism of evolution and to emphasise the central role of population genetics in this process. Certain universities have special interests and their courses reflect the particular interests of members of the department (see Ch. 6).

It is fair to say that there is now more concern at the tertiary level of education with teaching methods. Many universities now hold courses for their new staff on methods of teaching. With the rate of increase in knowledge there has been a tendency to include more material in degree courses and to emphasise the content above all else. This, together with some conflict of aims, has meant that some courses seemed to lack cohesion either because they appeared to some students to consist of isolated and disparate parts or because there seemed little connection between the theoretical and practical aspects of the work. These problems have been recognised and in many cases are being resolved.

Population genetics in schools
With a unified approach to biology well established in schools, the study of population genetics at the sixth-form level is not only valuable because of its key role in the study of evolution, but also because it links aspects of biology that have too often been divorced. Genetics, evolution, ecology and statistics are brought together to the benefit and advancement of all four. The BSCS projects, the Nuffield

Foundation—sponsored biology projects of the middle and late sixties, and A-level examination syllabuses give consideration to aspects of population genetics.

The well-known use of model populations, in which coloured beads or dyed peas are used to represent different alleles, is introduced pre-O-level in the first edition of the Nuffield O-level biology course. Their use is continued in the advanced level course where they are used to illustrate: no selection, strong selection in large and small populations, and the principles of genetic drift. With more schools having computer terminals, the natural extension of this work is in this direction where many generations can be worked through in a short time and where the computer may be programmed with a variety of data representing different selection pressures, mutation rates, and entry of new genes into the population (gene-flow) due to immigration.

Work is not confined to model populations at sixth-form level. Studies using *Drosophila melanogaster* are carried out, for example, setting up populations with equal numbers of wild-type flies and the *vestigial* mutant and recording the numbers of vestigial mutants at regular intervals.

It would also be possible to devise other experiments, for example, introducing a single wild-type fly into a small colony of the *ebony* mutant and recording the proportion of wild-type flies in succeeding generations, or to study (see Thoday and Gibson, 1962) the effect of disruptive selection using sternopleural chaeta number.

Work on natural populations is also undertaken in modern sixth-form courses and this, in my opinion, is the most profitable and interesting way of studying population genetics. Cyanogenesis in white clover, banding and background shell colour in the snails *Cepaea memoralis* or *Cepaea hortensis*, the style length in primroses, the numbers of *Corixids* in a pond over a period of years, or of grasshoppers in a given area, are all examples of studies which involve opportunities for personal observation and thus provide stimuli which provoke questions as well as attempts to find explanations based on hypothesis and experiment. They also lead to analysis of the results of these experiments, questions about the validity of the explanations, and, usually, to further investigation in the field.

Work of this type recognises no boundaries. It is imaginative, creative and integrative. It enables us to appreciate that evolution is a dynamic process and that changes in space and time can be investigated. There is no shortage of organisms for study and opportunities for original work abound.

References

R. Creed (1970) 'Short term studies in Evolution' in *Research Work in Colleges*, C. Smith and D. Adams, (eds).

W. H. Dowdeswell (1963 3rd edition) *The Mechanism of Evolution*, Heinemann.

W. H. Dowdeswell (1971), Ecological Genetics and Biology Teaching, in *Ecological Genetics and Evolution*, R. Creed, (ed), Blackwell.

E. B. Ford (1964) *Ecological Genetics*, Methuen.

E. B. Ford (1965 8th edition) *Mendelism and Evolution*, Methuen.

J. H. Gray (ed.) (1971) Nuffield Advanced Science, Biological Science, *Organisms and Populations*, A Laboratory Guide, Penguin.

H. B. D. Kettlewell (1958) 'A Survey of the Frequencies of *Biston betularia* (L) (Lep) and its melanic forms in Great Britain', *Heredity* 12.

M. J. Pentz (ed) (1971) *The Open University Science Foundation Course Unit 19*, Open University Press.

J. M. Savage (1963) *Evolution*, Holt, Rinehart and Winston.

P. M. Sheppard (1967 3rd edn) *Natural Selection and Heredity*, Hutchinson.

J. M. Thoday and J. B. Gibson (1962) 'Isolation by disruptive selection', *Nature*, 193.

Ecology and conservation

WILFRID DOWDESWELL

The scope of ecology defies precise delineation, but as our knowledge of biology advances it is becoming increasingly clear that its position within the total edifice is even more central than was supposed, say fifteen years ago. Part of the problem of definition derives from Haeckel's equivocal origin of the word ecology from the Greek *oikos*—house, *logos*— discourse; or, as he put it, the study of 'nature's household' (Haeckel 1866). Such a definition is not all that far removed from, 'the study of the structure and function of nature' (Odum 1963), which would seem to embrace virtually the whole field of biology. Others have equated ecology more specifically with the narrower area of field studies and hence with 'science out of doors' (Perrott *et al*, 1963). However, as we shall see later, such a definition is fast becoming outmoded in view of the numerous ecological studies now being conducted in the laboratory using microorganisms or models of larger ecosystems. So perhaps the traditional definition of ecology still remains the most acceptable, namely, 'the study of living organisms in relation to their environment' (Dowdeswell 1966), and it is this interpretation that will be adopted for the remainder of the chapter.

Small wonder that a subject with boundaries as diffuse and far reaching as ecology should present such unique problems for the teacher at all levels. The situation has been well summarised (Lambert and Goodman 1966) in a symposium of the British Ecological Society—'some think of it (ecology) as the easiest and most rewarding path of entry into the study of biology; while others regard it as too difficult and too complex to be tackled at all without considerable foreknowledge of other natural phenomena and experience of scientific method'. Clearly, between these two poles of thought there is plenty of space for manoeuvre.

Recent advances in ecology
In such a vast and amorphous subject, it is not easy to summarise in a few lines the main trends of development over the last decade or so. The following list therefore provides a basis for debate rather than any complete statement of fact:

1 *The development of more precise methods of measuring the physical environment* Recent advances have been brought about not only by the advent of more sophisticated electronic circuitry, but also through the increasing utilisation of such devices as transistors, photo-resistors, thermistors and oxygen electrodes. As a result, we now possess sensitive equipment for making continuous and accurate records of such physical phenomena as thermoclines and solar energy. Its advent has also opened up new and exciting possibilities for the study of microclimates (Southwood 1966; Wadsworth 1968).

2 *A better understanding of the relationship between the ecological requirements of organisms and other aspects of their make-up, such as their physiology and genetics* A good example of the former is provided by the extensive studies of heavy metal tolerance evolved by plants growing in the vicinity of high concentrations such as mines. It now appears that this phenomenon is world-wide and of great significance in influencing the local distribution of certain species, particularly grasses (Bradshaw 1970). Allied to such findings is the developing subject of ecological genetics which seeks to relate ecological and genetic approaches in terms of such features as survival rates and selection pressures. Investigations along these lines based on plants such as cyanogenic clovers, insects, molluscs and a great variety of other organisms have opened up a fresh dimension in ecology as well as posing many new problems for the teacher (Ford 1971).

3 *Improved quantitative methods and the development of new mathematical techniques* Significant advances in this field have led to an improved understanding of such aspects of ecology as spatial distri-

bution, competition between and within species, succession and energy flow, to mention only a few. Among animals, one area in which outstanding changes have taken place is in the estimation of numbers using the technique of mark–release–recapture, the effectiveness of which has been greatly enhanced by the introduction of new mathematical treatments, many involving computer programmes (Cormack 1968).

4 *The development of models to simulate ecological situations* The realisation that ecosystems are usually far more complex than they appear has prompted the need for simulation in order that one variable may be controlled while the effect of another is investigated. Such models may be purely mathematical and lend themselves readily to computerisation (Yarranton 1971). Alternatively, they may involve the study of living organisms under miniaturised conditions providing opportunities not only for original research, but also for new approaches to class teaching (Vogel and Ewel 1972).

5 *An increased knowledge of the relationships between the organisms constituting an ecosystem* Studies by Odum, Margalef and others have been prompted by the realisation that the flow of energy provides the unifying element within an ecosystem. They have also served to highlight the limitations inherent in measuring populations by traditional parameters such as numbers and biomass. The study of ecological energetics has thus provided a deeper insight into basic ecological theory and also paved the way for economic advances, such as greater crop productivity and the alleviation of world food shortage (Phillipson 1966).

6 *The application of ecological principles to the solution of environmental problems such as pollution* The study of problems in conservation encompasses much of ecological thinking—the structure of ecosystems, plant/animal interactions, the balance of physical and biotic environments, and the nature of adaptation by living organisms. Problems such as eutrophication (the loading

of waters with an excess of nutrient salts) are relatively new phenomena, and their origin and eventual cure are matters with which the ecologist is intimately concerned (Mellanby 1972).

Ecology teaching in Universities

It is difficult to generalise about undergraduate ecology courses in universities for these vary considerably depending upon the number of students, their seniority and the coverage required. In situations where an initial introductory course is provided covering basic ecological principles, a more advanced and specialised treatment may be possible later on. Sometimes, as at Bath, these two stages have to be merged into one, a situation which generates considerable problems in achieving an appropriate level of sophistication while ensuring a mastery of essential elementary principles. Experience over several years has shown, alas, that we cannot rely on this basic groundwork having been covered adequately at school.

However we care to define ecology, the fact remains that it is an integrated subject covering the interactions of plants and animals with their environment. While at an advanced level it may be desirable to split up ecosystems for convenience of study into their plant and animal components, when introducing the subject to students it is essential that it should be treated in an integrated way. In organisational terms, this may well necessitate two or more members from different departments (e.g. botany and zoology) cooperating in the running of a single course in which both are equally involved, and where both participate in discussions with students and in laboratory work. We are not concerned here with the detailed structure of a university ecology course at any particular level but rather to establish basic principles and procedures that need to be considered at some time or other.

During the last few years, considerable thought has been given at Bath to the design of an ecology curriculum (Dowdeswell and Potter 1973). The structure of our course centres round three main themes:

1 the spatial distribution of organisms
2 the flow of energy in ecosystems
3 the gene as an element of continuity in populations.

Besides serving as convenient reference points for discussions with students, these three themes provide a useful integrating element linking ecology with other areas of biology, such as physiology and genetics. Besides covering certain areas mentioned earlier where notable advances in ecological knowledge have recently been made, emphasis in the course is also placed on a number of basic ideas, some of which are treated from a teaching viewpoint in an unconventional way. The principal topics include:

1 Ecological terminology and the concept of species.
2 Spatial distribution and the factors concerned.
3 The physical environment and methods of measuring it.
4 Types of environmental change and their effects on distribution.
5 The relationship between distribution and physiology. Evidence from research on lampreys (Hardisty and Potter 1971).
6 Inter-relationships of organisms: predatory/prey, epibionts, commensals, symbionts and parasites.
7 Numbers of organisms, density dependent and independent factors.
8 Competition leading to succession and climax.
9 Structure of ecosystems. Energy flow.
10 Genes in populations, genetic polymorphism and selection.
11 Evidence for the action of selection derived from studies of ecological genetics.
12 Movement in populations. Dispersal in plants and migration in animals.

Most of these topics are covered by integrating lectures, seminars and small discussion group work with laboratory sessions and field studies.

An example of unconventional treatment is provided by the introduction of population genetics and selection in an ecological context as part of the study of populations. For practical work, the Hardy-Weinberg principle is applied to instances of selection and drift using bead models, while the influence of selection is studied through investigations of predation by the garden snail, *Helix aspersa*, on cyanogenic and acyanogenic birdsfoot trefoil, *Lotus corniculatus* (Darlington and Bradshaw 1966).

The objectives of our ecology course can be summarised briefly as follows:

1 to provide students with detailed practical experience of at least one ecosystem (in our case, a stream);
2 through lectures, discussions and further reading, to introduce other habitats treated more superficially;
3 through experience gained in (1) and (2) to enable students to gain an understanding of basic ecological principles.
4 to familiarise students with some basic ecological skills such as measuring physical factors, the construction of models in biological investigation and the use of certain statistical and sampling methods;
5 to illustrate the kinds of procedure involved in ecological research by using actual examples;
6 to show how the study of a series of research papers can throw light on the way in which the understanding of an ecological subject has changed (Epstein 1970). One topic chosen for study has been the incidence of cyanogenesis in white clover, *Trifolium repens* and birdsfoot trefoil, *Lotus corniculatus*;
7 to present ecology as an interdisciplinary subject having close links with kindred areas of biology, particularly physiology and genetics (Dowdeswell 1971).

The course at Bath was timetabled for one day a week with a duration of 10 weeks, provision being made for two one-hour lecture/discussion sessions in the morning followed by a two and a half hour laboratory period in the afternoon. Inevitably, much detail has had to be omitted from the summary above, but this is relatively unimportant as the same objectives could certainly have been achieved using a different range of topics and examples.

As was pointed out earlier, one of the biggest problems in designing an ecology course at university level arises from the need to include sufficient introductory material while pitching the bulk of the work at an intellectual level appropriate to the seniority of the students. Our scheme is something of a hybrid in that it attempts to meet both needs at once. The eventual intention is that it should be simplified and shortened into a truly introductory course, but one which can be elaborated and expanded to cater for more advanced requirements.

Ecology teaching in Secondary Schools
An advantage of schools over universities is that learning is spread over a far longer period of time. Ideas and attitudes introduced early on can thus be revived and elaborated at intervals to keep pace with a child's conceptual and intellectual development. This is particularly important in a subject such as ecology where the method of approach is as significant as the subject matter itself. The teaching cycle can be divided conveniently into three phases (Dowdeswell 1967) which apply irrespective of the range of ability concerned, although the depth of treatment will naturally vary from one situation to another.

Phase 1   (age range 11–13 years)
This is the introductory stage when treatment needs to be broad and simple, with emphasis on the individual organism and populations rather than the more complex interactions within communities.

Curricula at this level sometimes include biology as a separate subject, but more often they represent some form of combined science, usually with a fairly strong biology component. From an ecological standpoint, some of the principal objectives of such a course are to enable children:

1 to make simple observations and records;
2 to develop enthusiasm for and enjoyment of the subject;
3 to experience a wide range of living organisms and their life cycles, and to understand how they are named, identified and classified;
4 to have at least one opportunity of carrying out an investigation involving a living organism;
5 to appreciate through laboratory studies (e.g. of micro-organisms) that ecological principles apply in many

human situations, e.g. disease, souring of milk etc.

An ever-increasing range of organisms is now obtainable for laboratory study in schools. Among flowering plants a great variety of seeds is available, so there is no longer any excuse for using the broad bean to the exclusion of all else. The mung bean is now stocked by most suppliers and has proved particularly valuable on account of its rapid growth rate. Among animals, locusts provide the insect example *par excellence*, while earthworms and cryptozoic species, such as woodlice and centipedes, are good material for conducting elementary investigations in the field on such aspects as distribution and behaviour. Some teachers have been more ambitious in encouraging children to set up hypotheses explaining distribution observed in the field and to test these in the laboratory using equipment such as choice chambers. While such studies can be successful in the hands of able pupils, they demand a fairly sophisticated approach and are therefore best introduced at a later stage. Finally, a word concerning the introduction of the formal Linnean system of classification—as opposed to informal methods of grouping and naming living organisms. At one time it was supposed that the introduction of classificatory groupings above the species level provided an appropriate introduction to biology by bringing together at the outset a wide and diverse assemblage of plants and animals. However, experience has shown that such knowledge is best acquired gradually by experience over the two-year period, and that formal classificatory procedure provides a more fitting end point to the course than a beginning.

Phase 2   (age range 13—16 years)
As children grow older their fund of experience widens and they are able to view scientific problems in a more objective and analytical way. From the third year onwards the approach can therefore be increasingly broader, deeper and more quantitative. An early realisation of the fact that ecology begins on our own front doorstep—or even inside it—is important, and wherever possible ecological studies should be conducted in the vicinity of the school. At the end of such a course, which will usually occupy a small part of a longer one in biology or general science, pupils can

reasonably be expected:

1 to appreciate the significance of elementary qualitative and quantitative methods of study e.g. sampling of populations, the use of choice chambers etc;

2 to have studied spatial distribution in at least one population or community;

3 to understand the meaning of physical factors, and to have attempted to relate the influence of at least one of these to the situation studied in (2) above;

4 to realise the importance of behavioural mechanisms in influencing the distribution of animals, and to have studied at least one such mechanism in the laboratory e.g. in a choice chamber;

5 to be aware of some of the methods of dispersal and colonisation used by plants and animals, and to have studied a few at first hand;

6 to have encountered examples illustrating the range of relationships existing among living organisms— predator/prey, epibionts, symbionts, commensals and parasites;

7 to appreciate that plants and animals interact with one another and their physical environment, forming eco-systems based on producers and consumers; that such economic principles apply not only to wild communities but to human societies as well, and underline the solution to such vital world problems as pollution and conservation.

Throughout the course, children should become increasingly aware of the problems posed by ecological situations where many variables are involved, some of which may not yet be identifiable. This further serves to underline the importance of studying the effects of at least one factor under laboratory conditions (see 4 above) and of attempting to relate the findings to the natural situation. Throughout the three-year span, many opportunities will occur of gauging the relevance of ecological principles to the problems facing human societies—overpopulation, food shortage, conservation of resources and pollution, to mention only a few. Digressions into some of these fields will be well worth while, if only to demonstrate that ecology is not just an academic discipline but has an important applied side as well.

Phase 3 (Sixth form: age range 16—18 years)
Compared with the lower and middle school, the situation in sixth forms allows greatly increased scope for the study of ecology due to a greater allowance of time, the maturity of the students, smaller classes and better facilities. At its highest level, such a course can thus approximate in its general aims and content to an introductory course at a university, discussed earlier. Nowadays, it is customary in some schools to confine much of sixth form ecological work to a stay at one of the many excellent field stations, such as those run by the Field Studies Council (Sinker 1967). These centres are located in outstanding ecological localities, are well equipped, have expert staffs, and provide unrivalled facilities for practical classwork and the pursuit of individual projects (Tricker and Dowdeswell 1970). But their main purpose is not to teach elementary ecological principles— these must be acquired by pupils beforehand if they are to derive full benefit from the unique opportunities that a stay at a field centre has to offer. The following list of objectives covers a range of ecological skills and experience that a sixth former taking GCE A-level might reasonably be expected to have acquired during a two-year biology course. With mixed ability groupings now a feature of most sixth forms, it stands to reason that those with more modest aspirations will be able to cover proportionately less ground and at a lower level of sophistication. By the end of the course a student might be expected to have:

1 studied at first hand the species concept, including problems of identification and classification in plants and animals;
2 investigated the spatial distribution of organisms in at least one community, including a quantitative study of variations in density using appropriate sampling and elementary statistical methods;
3 understood the significance of adaptation in living things and to have studied one example practically e.g. the range of feeding mechanisms in organisms occupying a particular microhabitat;
4 studied the effect of a physical factor on the distri-

bution of a group of organisms and measured its range of variation;

5 worked out at least two links in a specific food chain using all available kinds of evidence;

6 appreciated the significance of the ecosystem concept, the balance of producers and consumers, and the assessment of communities in terms of numbers, biomass and energy flow;

7 realised the range of different relationships existing between organisms, and their significance in the biotic environment;

8 appreciated that the biotic environment is a dynamic system in a constant state of change due to such factors as competition and succession;

9 become aware of the range of variation, both genetic and environmental, exhibited by plants and animals, and studied at least one example quantitatively;

10 appreciated the magnitude of selection pressures acting on wild organisms, the nature of some selective agents and their mode of operation;

11 acquired a deeper understanding than previously (see Phase 2 (7)) of some applications of ecological principles to the solution of contemporary problems, such as pollution and conservation.

Phase 3 thus differs from the others not only in level of sophistication and factual coverage, but also in the opportunities it provides for diversifying methods of teaching and learning. For instance, an increased emphasis on the study of distribution, and the influence of physical factors upon it, has stimulated the development of simulation methods, such as the setting up of miniature static water communities of plants and molluscs maintained in large tanks (Gray 1970). These have proved an outstanding success and are preferable for study, in some respects, to the real thing. Other examples of the successful use of models in teaching occur in the field of succession (Darlington 1966), and inter- and intra-specific interaction (Gray 1970). Such developments are of great significance, particularly to urban schools, in helping to meet the basic needs of sixth formers in circumstances where facilities outside the school are limited or nonexistent.

Conservation

Although conservation forms part of the title of this chapter, I have detached it for separate treatment at the end, for a number of reasons. First there is the misconception frequently apparent in writings on this subject, that conservation (often erroneously confused with preservation) applies only to plant and animal communities. In fact its scope is far wider and covers the whole of the world's resources, including the atmosphere. Its study is thus essentially interdisciplinary, not just the province of biologists. A second misconception is that conservation can be taught as a topic in its own right, like algebra or electronics. My principal reason for omitting it from the earlier discussion is the belief that much of it grows naturally out of all that has gone before. Besides being an aspect of applied ecology, conservation, like ecology itself, derives ultimately from a way of thinking. The basic problem in teaching the subject is to present ecological thinking in a conservation context, with consideration not only for man's effect on the environment but also for the outcome of changes which take place naturally (Baron 1971).

In the three phases of secondary school ecology teaching mentioned earlier, much emphasis was placed on the study of environmental factors and their influence on distribution. Until comparatively recently such studies, even at sixth form level, have been confined to purely physical measurements of such factors as temperature, light and current, while chemical aspects have usually been restricted to oxygen and carbon dioxide. But a greatly increased awareness of the effects of pollution has now resulted in a wide range of new tests becoming available, so that the presence of potential pollutants such as sulphur dioxide in air and nitrate in water can readily be assessed (Andrews *et al* 1972). These methods are likely to exert a marked effect in the future on our teaching about physical factors, particularly in Phase 3 of secondary schools.

Another outcome of recent research on pollution has been the realisation that some species (indicator species) are more susceptible to pollutants than others. The principle is well illustrated (figure 1) by the small gastropod, Jenkins's spire snail (*Hydrobia jenkinsi*), a common inhabitant of freshwater streams. Snails kept in the laboratory in water polluted by

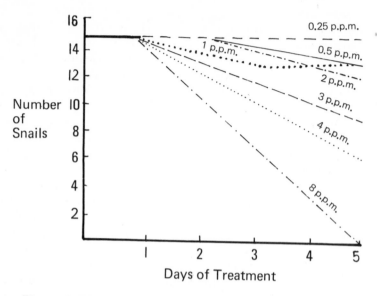

Figure 1 Effect of domestic detergent on the survival of Jenkins' spire snail (after Baron)

commercial detergent at a concentration of 0.25 ppm were unaffected after five days, while those in a concentration of 8 ppm all died. Animal and plant species can afford a sensitive means of monitoring pollution provided it is remembered that their populations are subject to periodic fluctuations in numbers as a natural outcome of ecological change.

While an understanding of the potential effects of pollution is important, it is also desirable that children should appreciate the dynamic nature of the environment, and that conservation and preservation are not synonymous. The pressures of selection, competition and succession are far more powerful than is often supposed, with the result that a natural habitat once isolated from them can rapidly become transformed. For instance, in the absence of grazing by sheep or rabbits, a piece of grassy chalk downland can be converted to impenetrable hawthorn scrub in the space of only five years. Conservation is an active process borne of ecological thinking. Bearing in mind its urgency for us today, perhaps this is one of the most powerful justifications for its inclusion in all school and university biology curricula.

References

W. A. Andrews *et al*, (1972) *Environmental Pollution*, Prentice-Hall.

W. M. N. Baron (1971) *Nature Conservatioh*, Methuen.

A. D. Bradshaw (1970) 'Plants and Industrial Waste', *Transactions and Proceedings of the Botanical Society of Edinburgh*, 41.

R. M. Cormack (1968) 'The statistics of capture — recapture methods', *Oceanography and Marine Biology*, Annual Review, 6.

A. Darlington (1966) In *Living Things in Action* (Nuffield O-level Biology, Year IV). Longmans/Penguin.

C. D. Darlington and A. D. Bradshaw (eds 1966) *Teaching Genetics*, Oliver and Boyd.

W. H. Dowdeswell (1966) *An Introduction to Animal Ecology*, Methuen.

W. H. Dowdeswell (1967) 'Ecology in the Nuffield Scheme', in *The Teaching of Ecology* (ed. J. M. Lambert), Blackwell.

W. H. Dowdeswell (1971) 'Ecological Genetics and Biology Teaching', in *Ecological Genetics and Evolution*, (ed. E. R. Creed), Blackwell.

W. H. Dowdeswell and I. C. Potter (1973) *An Approach to Ecology Teaching at University Level* (in press).

H. T. Epstein (1970) *A Strategy for Education*, Oxford.

E. B. Ford (3rd edition 1971) *Ecological Genetics*, Chapman and Hall.

J. H. Gray (ed. 1970) *Organisms and Populations*, (Nuffield Advanced Science), Penguin.

E. Haeckel (1866) *Generelle Morphologie der Organismen*, Reimer, Berlin.

M. W. Hardisty and I. C. Potter 'The behaviour, ecology and growth of larval lampreys', *The Biology of Lampreys* 1, Academic Press.

K. Mellanby (1972) *The Biology of Pollution*, Edward Arnold.

E. P. Odum (1963) *Ecology*, Holt, Rinehart and Winston.

E. Perrott *et al*, (1963) *Science Out of Doors*, Longmans.

J. Phillipson (1966) *Ecological Energetics*, Edward Arnold.

C. A. Sinker (1967) 'The Role of Field Centres in Ecological Teaching', in *The Teaching of Ecology* (ed. J. M. Lambert), Blackwell.

T. R. E. Southwood (1966) *Ecological Methods*, Methuen.

B. J. K. Tricker and W. H. Dowdeswell (1970) *Projects in Biological Science* (Nuffield Advanced Science), Penguin.

S. Vogel and K. C. Ewel (1972) *A Model Menagerie: Laboratory Studies about Living Systems*, Addison-Wesley.

R. M. Wadsworth (ed) (1968) *The Measurement of Environmental Factors in Terrestrial Ecology*, Blackwell.

G. A. Yarranton (1971) 'Mathematical representations and models in plant ecology' *Journal of Ecology* 59.

Further Reading

W. D. Billings (1972) *Plants, Man and the Ecosystem*, Macmillan.

J. Brierley (1969) *A Natural History of Man*, Heinemann.

M. Graham (1973) *A Natural Ecology*, Manchester University Press.

H. F. Hartmann *et al*, (1972) *Nature in the Balance*, Heinemann.

K. Mellanby (1972) *The Biology of Pollution*, Edward Arnold.

National Association for Environmental Education (1973) *Environmental Education*, vols 1 and 2, Heinemann.

The new medicine and the teaching of biology

RICHARD JOSKE

Medicine is a part of biology, applied human biology, which has both contributed to, and gained from, the new movements in biology. The consequences of this interaction are profound and contradictory. Medical practice has become more complex, technical, precise and costly, but at the same time moral and practical issues have arisen which require understanding and free discussion within the community. This chapter considers some causes and effects of these, and their relation to the teaching of biology and medicine.

Classical medicine has over the centuries been spectacularly effective. The measure of this is the dramatic increase in the world population which has risen from an estimated five millions in 8000 BC to some 4,000 millions at present, and which will probably double again in another generation.

The increase is due to several factors. There is little evidence of an overall change in human fertility, although the time of onset of puberty has fallen in most developed countries. There have been a major reduction in infant mortality and a great lengthening of the life span. The mortality rate of infants under one year in the United Kingdom in the middle of the nineteenth century was about 150 per 1,000 live births. The present figure is below twenty in countries such as Australia and the United Kingdom, although still over 100 in some parts of Africa and South America. The mean life expectancy of primitive man has been estimated at about twenty-five years. It approached fifty years in advanced countries at the beginning of this century, and now exceeds seventy years in such places as Scandinavia and the United States of America.

The medical and social reasons for these remarkable decreases in the death rate are not simple. They include public health measures, improved nutrition, and lowered mortality from famine, natural catastrophes, and infectious diseases.

Public health measures

Public health measures are quantitatively the most significant of these, but the simplest in conception, and with a long history, from the elegant treatises of Hippocrates to the multitudinous by-laws of a modern bureaucratic state. Public health measures aim simply to prevent disease by preventing the conditions that cause it. They include provision of clean air and water, sanitation, protection from noise, inspection of foods and food-processing, vaccination and immunisation against infectious disease, quarantine services, and industrial safety measures in mines and factories, among many other activities.

Many of these measures are not medical in the strict formal sense, but belong to domains such as engineering, architecture, town planning and politics. They are important in the present context, not only because of their past effectiveness, but because they represent a different tradition from that of classical medicine—care for the community as a whole, rather than for individuals. This can succeed in a democratic society only with general community support. It poses ethical problems, such as compulsion of some contrary to their beliefs, and in some instances, such as vaccination, actual danger to the lives of a few individuals for the potential benefit of the community as a whole.

Public health is a field that cannot be exhausted. Although many problems have been solved, new ones arise continually with increasing population density and urbanisation, new industrial processes such as the introduction of nuclear power, and new threats to the human environment, including loss of privacy, the effects of pharmaceuticals and household chemicals. This is an area in which project work by students can be peculiarly effective as a community weapon and method of education.

Nutrition and population increase

The problems of human nutrition appear at first sight to be solved. The normal dietary requirements of man are known and might, in theory, be provided, however poorly economic hardship and inadequate distribution fail to achieve this in practice. Once again, the rapid increase in human numbers is straining the biological resources of the ecosphere. The fate of many is still partial starvation, while in affluent countries new diseases are appearing as a result of overnutrition, such as obesity and some types of heart disease and diabetes. These problems also are of politics and economics rather than medicine, but an informed community seems a prerequisite for their solution.

Control of infectious disease appears an obvious success of traditional medicine, but has in fact been due largely to public health and preventive measures. Jenner showed the efficacy of vaccination with no concept of virus, and Pasteur developed the germ theory of disease from studies of fermentation in the brewing industry. The introduction of antibiotics has had comparatively little effect on the mortality from infectious disease. Present research is concerned mainly with development of new antibiotics, the emergence of resistant strains of organisms, and the occurrence of allergic reactions by patients to antibiotics. Study of infectious disease has led also to the study of biological warfare, a threat as dangerous as atomic warfare.

The population increase has led to changes of kind as well as number. This is shown best in the changing age structures of communities. A primitive tribe has a high conception rate, a high infant mortality and low life expectancy, resulting in a large proportion of young adults and few aged. A decreased death rate, especially of infants, and a longer life expectancy produces at first a population with a high proportion of children. This is seen to-day in countries such as Mauritius, where over 40 per cent of people are below the age of fourteen years. Such a population structure poses great problems in provision of services, for example, education for the large numbers of relatively unproductive young. It foreshadows a further increase in population when these children mature. Later, decreases in birth rate and further lengthening of the life span produce the population profile characteristic of long-developed countries, such as the United Kingdom, with a high and increasing proportion of the elderly and consequent further demand on social and welfare services.

A more subtle change, possibly of greater importance over

a span of generations, is an alteration in the human gene-pool. Many conditions are known which would be lethal in a primitive society, but which are compatible with both longevity and fertility when early diagnosis and continuing medical treatment are available. These include relatively common conditions such as diabetes as well as rare diseases such as phenylketonuria, which results in gross mental retardation if untreated. It has been suggested (with little evidence) that such a change in the genetic structure of man will be exacerbated by an increase in background radiation and mutagenic agents from atomic and industrial processes. Research into these problems takes several forms.

Population control is a complex and ultimately social issue. It might involve sterilisation, subsidised and freely available contraception and abortion, taxation increasing with family size, and changes in the mores of sexual behaviour and the meaning of marriage. Improved methods of contraception are also under intense study. For social acceptance such methods require altered social awareness, but all have been adopted in some part of the world. Their general use will require a general awareness of the possible social and ecological effects of overpopulation.

The nature of the ageing process is speculative, although genetic factors are certainly involved (Goldstein, 1971). It is doubtful if the potential life-span of man can be lengthened greatly, but within this limit much can be done by medical and social measures to maintain the health and well-being of the aged, and their ability to make a continuing effective contribution to the community (Busse, 1971).

Genetic engineering at the cellular level is impossible in man, and appears likely to remain so. Eugenic programmes require more consideration, but they are unlikely to be effective because of the low carrier and comparatively high mutation rates of many diseases, and because of social factors such as reluctance to interfere with the normal mating process and uncertainty as to the desirable qualities of man.

The primary results, therefore, of medical research are a rapid population increase and possibly also long-term changes in the human gene-pool. There are also secondary effects.

Modern living conditions

Most obvious are changes in the living conditions of a rising proportion of mankind. *Homo sapiens* evolved in Africa from small groups of territorial but ranging anthropoids. This is a way of life still characteristic of the small groups of hunting peoples left on earth, but which differs greatly from the congested urban habitat of most people to-day.

There is controversy whether this change of living conditions affects the health of mankind. It seems reasonable to assume that it does. Just as our physical development represents the result of organic evolution of our species, so mental development, which relates to the physical characteristics of the brain, may be similarly determined to greater or lesser degree by our evolutionary history. This is unproven but plausible, and exploration of this concept has led to the study of man in both physical and mental aspects as an evolved anthropoid species. This type of work is named ethology, and is the centre of much work (often uncritical and usually biased one way or another) in many schools of medicine, anthropology and social science.

There are certainly differences in morbidity and mortality between countries and between urban and rural areas of the same country. There is evidence suggesting that raised blood pressure and its consequences, such as strokes and heart disease, are more common in urban than pastoral societies, but even if this is so, the relative roles of genetic and environmental factors are uncertain. In England, cancer of the lung is more frequent in the towns than in the country, and this is almost certainly due to the greater pollution of the atmosphere in the former. Motor vehicle accidents are reaching epidemic proportions in urban societies.

Whether these patterns of physical illness are reflected in patterns of mental and psychological illness is a much more difficult question. There are no satisfactory criteria which may be used to compare one society with another. Suicide rates are to a large extent determined by cultural factors. Admissions to mental hospitals are evidence only of severe mental illness, and are affected by availability of mental health facilities, the health policies of government, and social usage. Criticism of society, acceptable in Western academic circles, may lead to admission to a mental hospital for 'deviationism' in other circumstances. However, study of psychiatric disability in African communities shows no basic differences from such illness in European society (German, 1972).

There is nevertheless some evidence that Westernisation of a rural society, and movement from a provincial to a metropolitan city are both associated with an increased frequency of psychotic breakdown (Hinkle, 1961; Murphy, 1961), unlike migration from one society to another similar one, as from Europe to North America. Although these results may be influenced by the more adequate health services in big cities, and the natural tendency of mentally ill persons to move from remote areas to places where more effective treatment may be expected, these are not the only factors. A survey in London has shown a significantly greater incidence of mental hospital admissions in boroughs within the flight paths of Heathrow airport compared with similar boroughs in other parts of the city.

Antisocial behaviour and crime rates also appear related to urbanisation, but again it is uncertain whether these are due to the psychological trauma of urban living, or to differences in opportunity for such behaviour and the difficulties of detection of offenders in the anonymity of a modern megapolis.

It is uncertain, therefore, whether reported differences in human behaviour, especially abnormal behaviour, associated with urbanisation are the direct results of urban living and its consequences, or whether they are secondary effects of psychological and possibly physical stresses leading to expression of disease which might otherwise be latent or compensated. For these reasons the study of man in an urban environment and under conditions with a high population density is one of the foremost problems facing medical research. It is beginning to be studied in many centres.

It is also being studied experimentally in animal societies under natural and artificial conditions and constraints. Some remarkable and chastening results have been published in recent years.

One of the best of these studies is that of Calhoun and his colleagues at the National Institute of Mental Health in Maryland (Calhoun, 1973).

They constructed for a group of mice a closed artificial universe with unlimited food supplies and absence of predators and disease. The colony began with four pairs of mice and the population grew, at first rapidly and then increasingly slowly, until it reached some 2,000 individuals living in conditions of considerable congestion after about 560 days. After this the number of animals fell with progressive rapidity until all mice were dead within 1,800 days and the colony extinct.

The extinction of the colony was due to changes in the behaviour of the mice when the population density was greatest, and they were continually jostled by their fellows. Males became unable to defend their nesting territory. Females assumed this aggressive role, rejecting and attacking the young, forcing them to leave their nests. The rejected young were unable to take part in normal social relations with their fellows, and their courtship and maternal behaviour was disrupted.

In the later stages of the colony the majority of females did not conceive, and such litters as did develop were abnormally small. Males reared under these conditions did not engage in reproductive behaviour, but were solitary, engaging only in feeding and grooming. These abnormal behaviour patterns were so engrained that such mice could not engage with their fellows and remained infertile, even when removed to a normal environment with normal mice.

These results cannot be extrapolated to urban man without qualification and caution, but Calhoun believes there is no logical reason why a comparable sequence should not apply to other species, including man.

It is apparent that the population increase and its consequences, social, economic, mental, physical and genetic, pose major problems for medical and biological research, as well as requiring an informed community for their investigation and such social action as is necessary for their remedy.

Modern medical research
The direction of classical medical research has also altered in recent years. The decline of malnutrition and many infectious diseases in developed societies has led to increased study of other types of disease. Some of these are rare.

Others, such as arteriosclerosis, are becoming more frequent as people live to greater ages; others are developing from changed conditions of life, such as motor vehicle accidents and cigarette and other drug addictions; others reflect the influence of basic biological sciences upon medicine, for example, cancer and immunology.

Immunology developed from the study of infectious disease, when it was observed that a person exposed to an infectious agent on a second occasion showed immunity to the agent or reacted differently. Thus an episode of mumps protects a child from further attacks of this disease. This observation is the basis of many preventive measures against infectious disease. The change in reactivity is due to the production by the body of circulating proteins (antibodies) which react with, and may nullify the effects of, the foreign substance (antigen), whether it is an organism or a more simple protein, such as tetanus toxin or blood group substance. There is also a change in the fixed defence cells of the body which is of particular importance in diseases such as tuberculosis.

Occasionally these immune mechanisms may have a harmful rather than a protective action. Asthma results from an antigen-antibody reaction to an inhaled foreign protein occurring within the bronchial tree.

Recently it has been found that these immune mechanisms are concerned also with the normal control of body cells. If a portion of the liver of an animal is removed, the remainder grows until the initial mass of liver tissue is restored. If abnormal cells arise in the body, in most instances immune reactions lead to their containment or rejection. Cancer is basically a condition in which some body cells mutate so that they grow progressively without respect for these normal limiting factors.

Less obvious forms of immunological abnormality are now thought to account for the progression of some forms of liver, kidney and other diseases, although they may be initiated by causes such as virus infection.

The capacity of some body cells (immunocytes) to recognise and respond to foreign protein is the basic biological problem of organ transplantation. All cells of any person are of the same immunological type, to which his (or her) immunocytes do not normally respond. Transplantation into the body of cells of a different type produces a rejection reaction leading in most instances to death of the transplanted cells. Successful organ transplantation therefore requires either the use of a carefully matched donor (such as an identical twin) or artificial suppression of the normal immune response. This has been largely accomplished, and kidney transplantation is now a standard procedure in most major medical centres.

It is nevertheless a somewhat restricted technique. There are four requirements: some means of maintaining the recipient until a donor organ is available, a supply of donor organs, an adequate surgical technique for transplantation, and evidence that the transplanted organ will function and maintain the life of the recipient.

These requirements are satisfied in renal transplantation, but not for most other organs. For example, the complexity of the nervous system and the failure of nervous cells to divide and grow in adults make brain transplantation impracticable in the foreseeable future.

The first requirement, a support system for the recipient until a donor organ is available, raises the greatest problems. Life may be sustained after removal of both kidneys by the use of an artificial kidney. This is possible because an artificial kidney is required only intermittently, two or three times weekly. But an artificial heart or lung is needed continuously, since an absent circulation for even a few minutes causes irreversible death of brain cells. Although much work is being done on the development of artificial organs, such as the heart, results to date are extremely disappointing.

A further problem is the selection of recipients for transplants and other procedures, since both medical and social criteria are involved. They include age, psychological factors, the ability of the recipient to continue treatment indefinitely, and economic and family circumstances, as well as the formal medical indications. Many of these involve community as well as medical responsibility.

The cost of modern medicine

The final major theme is that of finance. Medicine is an expensive industry which absorbs some 5 per cent of the national income. Costs are rising steadily. This is due to increasing hospital costs, more sophisticated medical care, the demand for a rising standard of medical attention, the increasing prevalence of chronic disease, and the numbers of the aged and infirm in society.

Till recently medicine has not been subjected to any form of cost-benefit analysis, but it is doubtful if this freedom can continue. Medical services will have to compete within the national budget against other social services, such as education and housing, as well as against armaments, payments to the United Nations, development of supersonic aircraft, alcohol, tobacco and gambling. Hence in the future there will have to be cost-benefit studies within medicine, and between medicine and other community activities.

On any such basis, many publicised developments in medicine are extremely expensive. Figures given by Leach (1972) suggest that an expenditure of £1,000,000 on artificial hearts (if these were available) would save thirty-eight lives for five years; if spent on artificial kidneys 105 lives for five years; on heart transplants 250 lives, and on a successful campaign for seat belts, 29,000 lives.

With a limited budget, priorities in medicine will therefore become increasingly important. These priorities will apply not only to research and development, but also to selection of types of therapy and of patients for treatment by established methods. They involve ethical decisions. Should an available donor kidney be used for an alcoholic of twenty years, or a productive senior public servant of fifty years? Should we develop and increase facilities for treating advanced heart disease in a few, or should this money be used to treat the larger number of people with chronic lung disease due to cigarette addiction, or for rehabilitation of victims of motor accidents, or for the investigation of criminal and antisocial behaviour?

These types of decision will be required more and more often, as the costs of medicine rise relatively faster than the national income, or at least, that portion of it available to the health services.

The theme of this essay is that past medical research has led to new problems, such as overpopulation and ageing and their consequences, and that these problems are in turn directing current medical research. In nearly all instances these new questions, if solvable, will have solutions which lie beyond the formal, traditional bounds of medicine. They will require social and legislative action, which in turn means decisions by people other than the medical profession. If these decisions are to meet the humanitarian tradition of Western civilization, the economic and social realities of our time, and constrain the few for the benefit of the many, an informed and educated body of public opinion is mandatory.

The teaching of biology

The final problem to consider is, therefore, how to introduce these topics into the educational system, not only in the profession of medicine, but more widely at tertiary and (desirably) secondary levels.

At tertiary level this is not difficult. Courses in human biology have been introduced successfully in some universities, and texts of varying scholarship and bias are appearing.

Human biology is also academically respectable. Knowledge of *Homo sapiens* is greater than knowledge of any other species. A course devoted to the study of man should therefore be acceptable both to the older purists, as providing study in depth, and to the newer fashionable academics, by providing a cross-disciplinary curriculum. The difficulty is to select from the large amount of available data in order to provide, firstly, a core of basic biological knowledge useful to specialists in zoology, psychology and cognate disciplines; secondly, some understanding of the biological and social nature of man; and thirdly, to introduce in a non-emotive fashion the social and political issues arising from these, and necessary for effective study in the social sciences. It is also desirable to provide a terminal course suitable for non-specialists taking a general degree, prior to careers such as teaching or the civil service.

The logical design for such a course would be to begin with cell structure and function, including the principles of Mendelian and polygenic inheritance, and then discuss the integration of cells into tissues and organs, and the association of these into multicellular organisms, including man. The principles of physiology and evolution could then be considered, and the relationships between organisms and their environment. This basic biology would be illustrated from the abundant human material available, such as inheritance of eye colour and finger prints, and examples such as the occurrence of haemophilia and porphyria in European monarchies. This leads naturally to the study of human variation and the races of man and of human ecology. The latter includes study of human nutrition, the differing environments of nomadic, agricultural and urban societies, including cultural adaptations, population structures and the differing causes of death under varying circumstances.

Concurrent with this formal course would be the deeper study of selected contemporary biosocial problems, some of which have already been indicated. They might include population growth and its control, especially in the light of changing sexual mores and of disease and natural disasters; the effects of noise, congestion and loss of privacy upon man, both in urban life and special circumstances, such as the imprisonment and brain-washing of political and military prisoners; the roles of aggression and competitiveness in sport, commerce, war and peace; the interaction of genetic and social factors in causing crime and antisocial behaviour; the physical and mental consequences of ageing; and the value of human life. Others could be mentioned. Selection will depend upon the issues of the time, the resources available, and the types and maturity of the student groups.

These topics would be studied as projects by small groups of students, to balance the more formal tutorial and laboratory teaching of the other part of the course. They would relate also to other courses such as zoology, psychology, anthropology, politics and social science.

This approach, combining basic biology orientated to man with project study of relevant social issues, is also practicable for senior school students. Potential biologists may require greater emphasis on the early part of the syllabus, including some discussion of the genetic code, and a quantitative approach to genetics and evolution, together with some experience of statistics and population and census data. The potential humanist may benefit from a more qualitative study emphasising broad bio-social principles.

Project work should again be the basic method of teaching. Much can be done through social agencies, such as hospitals, local councils and police, welfare and geriatric services. Students could survey noise, atmospheric pollution, land use, traffic problems and accidents in relation to place, types of car and driver, and racial factors in the local community.

This essay has suggested that progress in medicine has answered some questions, but produced new problems of increasing importance. These are of a type and magnitude that make them ultimately a community responsibility. Decision in medicine has formerly been restricted largely to a professional elite encompassed by tradition. Future decisions will require an open community where ethics, performance and priorities are the concern of all. This is practicable only if social aspects of medicine are considered as part of human biology, and studied widely in schools and universities.

References

E. W. Busse (1971) 'Biologic and Sociologic Changes Affecting Adaptation in Mid and Later Life', *Annals of Internal Medicine*, 75.

J. B. Calhoun (1973) 'Death Squared: The Explosive Growth and Demise of a Mouse Population,' *Proceedings of the Royal Society of Medicine*, 66.

G. A. German (1972) 'Aspects of Clinical Psychiatry in Sub-Saharan Africa', *British Journal of Psychiatry*, 121.

S. Goldstein (1971) 'The Biology of Ageing', *New England Journal of Medicine*, 285.

L. E. Hinkle (1961) 'Ecological Observations of the Relation of Physical Illness, Mental Illness, and the Social Environment', *Psychosomatic Medicine*, 23.

G. Leach (1972) *The Biocrats. Implications of Medical Progress*, Penguin Books.

H. B. M. Murphy (1961) 'Social Change and Mental Health', *Millbank Memorial Fund Quarterly*, 39.

Further Reading

The references cited above are generally suited for non-professional as well as medical reading. In addition, some recent works dealing with the major themes of this chapter which might form the basis of a course of study include:

F. M. Burnet and D. O. White (1972) *Natural History of Infectious Disease*, 4th edn., Cambridge University Press.

A. Chase (1971) *The Biological Imperatives*, Holt, Rinehart & Winston.

E. J. Clegg (1968) *The Study of Man. An Introduction to Human Biology*, English Universities Press.

P. R. Ehrlich and A. H. Ehrlich (1970) *Population, Resources, Environment*, Freeman.

G. Hardin (1972) *Exploring New Ethics for Survival*, Viking Press.

B. S. Hetzel (1971) *Life and Health in Australia*, The Boyer Lectures 1971, Australian Broadcasting Commission.

C. E. Johnson (ed.) (1970) *Human Biology: Contemporary Readings*, Van Nostrand Reinhold.

K. L. Jones, L. W. Shainberg and C. O. Byer (1971) *Age of Aquarius. Contemporary Bio-Social Issues*, Goodyear Publishing Company.

G. J. V. Nossal (1970) *Antibodies and Immunity*, Penguin.

L. Thomas, (1973) 'Guessing and Knowing. Reflections on the Science and Technology of Medicine', *Saturday Review (Science)*, 55.

World Health Organization (1972) *Health Hazards of the Human Environment*, World Health Organization.

J. Z. Young (1971) *An Introduction to the Study of Man*, Oxford University Press.

New approaches and techniques

Goals and objectives

DORIS FALK

In the last decade many changes have occurred in our society which have profound implications for science education in general and biology education in particular: the attitudes and motivations of young people, the enormous increase in the sheer amount of scientific knowledge, and new and challenging insights into the nature of learning. The encroachment of science into our daily lives and the increasing control that science can exert over nature makes it necessary for people to evaluate and make intelligent decisions about fundamental problems, such as the use of insecticides, population control, drugs, organ transplants, euthanasia, etc. These problems can be discussed most freely and objectively within the framework of a biology course. Biology teachers, therefore, have the responsibility to re-examine and modify their goals to make them more realistic in terms of the needs of young people in our present society.

DEFINITION OF GOALS

Goals in education are statements of the philosophical beliefs and assumptions upon which an instructional programme is founded. They convey the intent and determine the direction of that programme. The organisation and implementation of any programme, whether it be the programme of an entire school system, or the specific programme of one biology teacher, should be determined by the basic goals held by that administrator or teacher.

In order to see how goals function in the development of the overall teaching strategy, let us examine a few of the goals which a biology teacher might formulate for himself:

1 *To develop the knowledge of those biological facts, concepts, and generalisations which lead to an understanding of the unifying themes of biology* This

goal suggests the rationale of this teacher for the selection of the *subject matter* to be taught. We know that it is impossible to teach all there is to know about biology. We also know that research shows that a student can remember and use knowledge learned in a given situation only to the extent that he can relate it to the underlying principles which give structure to the discipline. The learner must understand the basic structure if knowledge is to be retained and transferred to the solution of new problems. Facts or isolated specifics should not be learned for their own sake but only as they contribute to the development of more permanent concepts, generalisations and themes. The unique characteristic of the knowledge (or products) of science is that the facts from which its concepts and generalisations are derived, are observable, measurable and testable facts about the phenomena of nature. This knowledge is generated through actual scientific investigations or experiments in which the investigator uses various skills, often called 'inquiry' skills or processes.
This leads us to a second goal our teacher might define.

2 *To develop the powers of critical observation and description, and the thinking processes which underlie scientific inquiry* This goal suggests that for this teacher, the 'processes' of science are as important for the student to learn as the 'products'. As mentioned above, the products are the result of scientific investigation and constitute what we think of as the subject matter of science. To know what science really is, students should also know how it works in arriving at these results. This teacher would provide the students with opportunities to engage in scientific inquiry: to formulate problems and hypotheses, to devise and perform experiments, and to predict and verify results.

3 *To develop those values and attitudes which are essential to scientific inquiry and underlie all rational thought, and thus guide future actions and relationships* As a scientist engages in the intellectual processes of scientific inquiry (described above), his progress is controlled and evaluated at each step of his

investigation by certain attitudes he knows are vital to the ultimate solution of his problem. Integrity, objectivity, open-mindedness, demand for verification, avoidance of dogmatism, cooperation, are all attitudes required of anyone, scientist or student, seeking answers to questions about nature. They are also related to basic questions which men ask about things totally unscientific—religious, aesthetic, humanistic and literary. Situations constantly arise in the classroom, and particularly in the laboratory, where choices must be made, results interpreted, explanations given, criticisms offered, conclusions drawn. The perceptive teacher will take advantage of such opportunities to encourage the growth and development of these positive attitudes.

USING GOALS

The number of goals could of course be expanded. In practice goals are interdependent and interrelated. For instance, it would be impossible to learn the processes of science in a vacuum, that is, in the absence of knowledge of scientific facts and concepts. And neither the products nor the processes can be learned in the absence of some set of values and attitudes.

Any statement of goals is subject to the interests, experience and philosophy of the individual teacher. That all teachers do not have the same perspective was recognised by the group of scientists and educators in the United States who developed the new, and now widely accepted, programme for secondary school biology, the Biological Sciences Curriculum Study (BSCS). They developed *three* biology textbooks and three sets of laboratory materials. One version is for teachers who prefer the ecological and behavioural approach to biology; another for those who feel that the understanding of the unity of all living things is clearest when approached from the biochemical or molecular viewpoint; the third is for those who find the meaning of biology is most apparent using the cell and individual organism, their reproduction and development, as the central focus.

The teachers who use these three versions may or may not have the same goals. If the goals are as broadly stated as those in the preceding pages, they could be the same, but the emphasis and interpretation would be different. They might all agree that broad rather than specific knowledge should be sought, and that the learning of scientific processes and attitudes are important goals; but the goals will be reached by different routes.

Whatever the goals of the teacher might be, the important thing is to *have* them and to *use* them. We are all too familiar with 'courses of study' and 'curriculum outlines' and 'lesson plans' which invariably begin with long lists of impressive-sounding goals, but more frequently than not, are never referred to again or implemented in the pages which follow.

DEFINITION OF INSTRUCTIONAL OBJECTIVES

Instructional objectives are descriptions of what the student should be able to do as a result of instruction. They describe intended *outcomes* rather than *content*. Objectives are derived from and are consistent with one's goals. They describe testable, observable behaviour which it is hoped will be acquired by the learner by the end of some specified instructional period. It is assumed that the cumulative effects of the students' achievement of the numerous objectives during the period of biology instruction will result in the attainment of the goals.

The teacher must decide in advance what he expects the student to learn, then state in behavioural, or performance, terms what the learner will be *doing* when demonstrating his achievement of the objective. Such objectives are also variously referred to as 'behavioural,' 'operational' or 'performance'.

Instructional objectives may be stated for long- or short-range learning. Defining long-range objectives, i.e., for a semester or year, gives unity and direction to the whole course. More intermediate objectives might be written for one unit or some portion of the whole course. The short-range objectives describe the instructional intent for a single lesson or lesson sequence.

RELATION OF OBJECTIVES TO GOALS

We might consider the broad goals as the horizontal dimension of the biology curriculum, whereas the instructional objectives constitute the vertical dimension. The objectives discussed in the pages to follow will be derived from the previously listed Goals 1 and 2. Stating instructional objectives related to Goal 3 would have little meaning. It is unrealistic to try to plan activities specifically for the development of 'honesty' or 'open-mindedness', for example. However, behaviour-changes can be identified which indicate progress towards the development of attitudes and values. For instance, in writing up results for laboratory experiments, if they do not come out as they should, the student would be expected to record what he finds, not what he thinks he was supposed to find. He would identify possible sources of experimental error and, if possible, repeat the experiment, using greater care. Such attitudes are encouraged in day-to-day activities and in many unexpected and unpredictable situations.

ADVANTAGES OF USING INSTRUCTIONAL OBJECTIVES

Stating instructional objectives is simply verbalising a description of observable performance, that is, what the teacher expects the student to be able to *do* or *say* or *write* after instruction, that he could not have done or said or written before this instruction. This aids instruction for both teacher and student in the following ways:

The teacher is forced to define clearly what he believes is important for his students to learn.

Specifications are provided for the selection of content, procedures, materials, and techniques.

Valid criteria for evaluating student achievement are provided by stating in advance the expected accomplishment.

The student is able to organise his own work and evaluate his own progress when he knows what is expected of him.

Classroom procedures which require active participation of students tend to be emphasised when objectives are stated in action terms.

INSTRUCTIONAL OBJECTIVES IN THE TEACHING STRATEGY

When a teacher begins to plan his teaching strategy, he goes through the following sequence of activities:

Long-range:

Decides on his major goals for the semester or year.

Makes an outline of his course or programme by selecting the topics and determining their sequence and organisation.

Short-range:

Prepares a set of instructional objectives for the unit and for the daily lesson under consideration.

Selects procedures, content and methods consistent with the objectives and principles of learning.

Provides for a means of evaluating the student's performance in terms of the objectives. (More about this later.)

It should be stressed here that only *after* the instructional objectives have been stated can the selection of classroom activities (laboratory work, discussions, demonstrations, films, etc.) and subject matter content, be made. 'You cannot concern yourself with the problem of selecting the most efficient route to your destination until you know what your destination is.' (Mager, 1962).

SELECTING INSTRUCTIONAL OBJECTIVES

Let us describe a teacher starting to plan a short unit on diffusion and osmosis. 'What do I think is important for my students to know about osmosis?' he asks himself. First, he takes into consideration his major goals: 'I don't want the students to memorise a lot of definitions (osmosis, diffusion, permeability, concentration, etc.), but I want them to understand the principles of diffusion and osmosis, since they will encounter them frequently in the future, and I want them to be able to apply them (Goal 1). Also, I am going to plan some laboratory experiences for this unit since it is possible for students to design some rather simple investigations to answer important questions about diffusion and osmosis (Goal 2).' What are some specific objectives this teacher might prepare on the subjects of osmosis and diffusion which would direct student-learning towards these goals? Is it possible to state these objectives in such a way that the teacher could tell if the objectives are achieved?

Suppose our teacher began with a comfortable and commonplace objective like this:

The student should develop an understanding of the process of osmosis.

This is a laudable objective but it doesn't provide much instructional help for the teacher or student. Let us take a closer look.

STATING INSTRUCTIONAL OBJECTIVES

An instructional objective must be stated so that the instructional intent is clearly communicated to the student. What should the student *understand* about osmosis? What should he be able to do to show that he *understands* osmosis? Does the teacher expect him to be able to recite, draw a diagram, define it? What kind of questions or problems or activities can the student be presented with that will provide the teacher with some evidence that the student does or does not *understand* osmosis?

Use behavioural terms

In stating the objective, avoid the use of terms or phrases which are too vague or open to a wide range of interpretations, and which do not specify what is to be learned, or the behaviour that will demonstrate whether or not it has been learned. The following are some examples to avoid:

develop an understanding of...	know
acquire an appreciation of...	understand
know the significance of...	recognise

Instead, words should be used which denote action or what the learner can *do* to show that he has 'an appreciation of...' or knows 'the significance of...'. The following words describe some kind of performance which will give evidence of 'knowing' or 'understanding' or 'recognising':

identify	draw
state in your own words	tabulate
describe	interpret
demonstrate	estimate
classify	suggest a procedure
construct	formulate an hypothesis
select	predict what would happen if...

These words describe *behaviour* so that the student knows what he is expected to do and the teacher knows what he will look for.

Let us see now what objectives our teacher might prepare that would incorporate this idea of using terms that describe the desired behaviour by name. By the end of this lesson the student should be able to:

1 *Construct*, when given an assortment of materials, an experiment to illustrate osmosis. (Derived from Goal 2)
2 *Predict* what would happen to a lettuce leaf if it were transferred from fresh to salt water. (Derived from Goal 2.)
3 *Draw* an arrow to indicate the direction of water-flow when given a diagram of two solutions of differing concentrations separated by a semi-permeable membrane. (Derived from Goal 1.)
4 *Demonstrate* 'permeability' and 'impermeability' with reference to the wire mesh, when given a piece of wire mesh, some sand and some beans. (Derived from Goal 1.)

Specify conditions for minimum performance
Whenever possible, describe the conditions that will constitute acceptable performance. The following are some examples of objectives in which such conditions are specified:

Define *in your own words* the meaning of the term 'solution'.
List *at least three examples* of the operation of osmosis in living organisms.
Given a list of situations in which diffusion is occurring, select *the three* that represent osmosis.

Without looking at your laboratory notebook, draw a diagram of the apparatus used in the osmosis demonstration.

In these last examples, not only are the desired behaviours specified (define, list, select, draw) as in the examples in the immediately preceding section, but in addition, the more explicit details (the italicised portions) are added to give the student a clearer idea of what is expected of him.

Specifying conditions in this manner does not apply to all statements of objectives. In the first four examples further conditions are not necessary to make the intent of the objective clear.

Represents a variety of cognitive levels

At any given time in any given classroom, students will vary widely in their intellectual development or 'cognitive' level. More specifically, this means that the understanding of any concept being discussed in class, e.g., osmosis, will vary from student to student, depending on their intellectual and physical abilities, environment, past experience and training. Each individual has a 'cognitive structure' (Bloom, 1956), the term used for the hierarchy of categories into which the cognitive (intellectual) skills possessed by each individual may be classified. The rate and level of learning new information for each individual depend upon his cognitive structure. Briefly, the levels of the cognitive structure are:

Knowledge: The learner can recall specific facts, or simply bring to mind the appropriate material, to memorise, remember.

Comprehension: The learner can compare, reorder or rearrange learned material, for instance, by explaining in his own words, giving examples, giving similarities and differences.

Application: The learner can apply concepts and generalisations he has learned to new, practical, or unfamiliar situations.

Analysis: The learner, when presented with an organised whole or structured situation, can break it down and identify its elements, as, for instance, when presented with an experimental design, he can identify variables, controls, hypotheses.

Synthesis: The learner can put elements or parts together to form a whole, a new structure or pattern. For instance, the student is able to design an experiment or propose ways of testing hypotheses.

Evaluation: The learner can make judgments about the value or accuracy of presented materials against established criteria. For instance, he evaluates proposals or hypotheses of persons in authority, in terms of probable results.

Knowledge, Comprehension, and Application are the foundation skills which are necessary for the assimilation of the skills of Analysis, Synthesis, and Evaluation. Each succeeding level represents a higher level of understanding. If the student can operate at the Analysis level, for instance, it is assumed he understands and could perform tasks at the levels of Knowledge, Comprehension, and Application.

Let us see how this might be applied by our teacher in preparing his instructional objectives. By requiring thinking at higher cognitive levels in some of his objectives, he would provide the incentive for his abler students to attain their highest level of understanding. At the same time, he would want to provide some tasks that *all* the students in the class could successfully perform. For those students whose cognitive structures are not so highly developed, some of the objectives would be prepared for lower levels of understanding. By preparing a relatively large number of instructional objectives for each unit at different levels of 'difficulty' (according to the categories of the cognitive structure), provision is made for the wide range of individual differences represented in the class.

The statements of the instructional objectives determine the level at which the student will be expected to operate. Most of us are guilty of not challenging our students enough. We present them with tasks and ask them questions that require no more of them than ability to recall or make minor reorganisations of factual material (the lowest cognitive level).

If you were asked to write out some objectives for tomorrow's lesson, wouldn't many of them look something like these?

The student should be able to:
   State the four elements found in living matter.
   Name the structures of the human heart.
   List three factors necessary for photosynthesis.

To be sure, some objectives at this first level must be included. But a conscious effort must be made to include objectives requiring the skills of the higher cognitive levels. This is not easy. Instructional objectives which require recall only are by far the easiest to prepare. The temptation is great to confine objectives (and test questions) to this category.

Some examples of instructional objectives at the different cognitive levels are as follows:

The student should be able to:
   Level 1 (Knowledge) Define the terms osmosis and diffusion.
   Level 2 (Comprehension) State in what way osmosis and diffusion are (a) alike (b) different.
   Level 3 (Application) Explain what would happen to a jellyfish living in the ocean, if it were placed in a bucket of fresh water.
   Level 4 (Analysis) Identify the factor in the osmosis demonstration that must be changed in order to change the direction of the flow of the water molecules. (This could be an example taken from the classroom experience or elsewhere.)
   Level 5 (Synthesis) Suggest a hypothesis to account for the fact that the roots of a plant shrivel and dry when a highly concentrated solution of fertiliser is applied to the soil.
   Level 6 (Evaluation) Find the fallacy in a doctor's report on his treatment of a patient who dies, but the doctor fails to account for osmosis in his explanation of death. (The treatment involves osmotic phenomena, such as injecting saline solution.)

Turn back to page 146 and review the examples of instructional objectives presented there. See if you can classify them according to level. You will find that it is not always possible to classify them at one level only. It can be argued that many of them fit into two or even more categories. This is really not so important. The value of using the idea of these different levels in preparing your objectives is that it makes you try to think of ways in which you can make your students *think* and *use* their knowledge, rather than simply regurgitate memorised definitions, unrelated facts, and other isolated bits and pieces of information that they have read in the book or copied from the blackboard.

USING INSTRUCTIONAL OBJECTIVES

Students will benefit most from instructional objectives if they know what they are at the beginning of an instructional unit. For some obscure reason, teachers are generally reluctant to give them their objectives in advance. Perhaps they are afraid that the students will learn just what is in the objectives and no more. If there are a number of objectives, if they represent what the teacher thinks is important in a given unit, and if they require the kinds of critical thinking described in the previous section, then what are the objections if the students study 'just for the objectives'? Instructional objectives are not limited to just what is done in class. They may ask the student to be able to apply a principle learned in class to some entirely new situation, or to be able to suggest some new way to design an experiment.

The objectives can be written on the blackboard, or they can be duplicated and handed to the students at the beginning of a unit or a lesson. With the instructional objectives as learning guidelines, more student independence and less teacher 'telling' can be expected.

INSTRUCTIONAL OBJECTIVES IN EVALUATION

Any kind of evaluation must be made in terms of certain pre-established standards or criteria, as, for instance, an automobile is evaluated in terms of how it performs with respect to horse power, petrol consumption, etc. When we evaluate student achievement we do it in terms of how the student performs with respect to some pre-determined criteria. For instance, we would not evaluate a student's laboratory report on the basis of artistic merit unless this was one of the specified criteria known to the students in advance.

The criteria for evaluating student achievement are the

long-range goals and the shorter-range instructional objectives established prior to the evaluation. One of the most important principles of teaching is that *evaluation must be in terms of stated goals and objectives*; otherwise, it is irrelevant, unfair, and useless.

If your selection of instructional objectives represents what you believe is most important for your students to know in a given instructional unit, then these same instructional objectives should form the basis of your evaluation of student achievement, both in daily performance and in examinations. This does not mean that test questions must be the same as instructional objectives. They should, however, be related to them in that they reflect the student's ability to perform the same skills and demonstrate his understanding of the same principles as those required in the instructional objectives. If instructional objectives require process or inquiry skills (hypothesising, interpreting, predicting, etc.), then there should be test questions requiring these skills.

The writing of test questions is greatly simplified if instructional objectives have been prepared. Review the examples of instructional objectives on pages 146 and 149 and it will be apparent how easily test questions can be derived from them. All cognitive levels should be represented in test questions, just as they are in the instructional objectives. Test questions requiring other than simple recall are difficult to formulate, but if wanting students to *think* rather than *memorise* is one of your serious goals, then it is worth the effort to develop the technique of writing this kind of question (Falk, 1971)

The instructional objectives also provide the means for continuous daily evaluation of student performance by the teacher, and the student has the means to organise his activities and to evaluate his own progress along the way. His success is not dependent on his ability to psychoanalyse his teacher in an attempt to predict what would impress the teacher or what the teacher might think was important. Stated more positively, the student's time could be spent on completing a set of clearly-stated tasks, the accomplishment of which he knows, with confidence, will mean success.

SUMMARY

It is no longer possible to teach all there is to know about biology. The annual increase in the sheer amount of biological knowledge staggers the imagination. Any biology course is no more than a teacher's small selection from this vast field, based on that teacher's ideas about what is important to know about biology.

Since a selection must be made, the teacher must have reasons for selecting what he does. The major selections are made according to his long-range goals. The more immediate selections of what the students are expected to learn, and the specifications for the conditions for learning, are spelled out in the statements of the unit or daily instructional objectives. These objectives are stated so that the student knows what he is expected to do and his performance can be observed and evaluated by the teacher. Thus, the instructional objectives are simultaneously the blue-print for learning and for evaluation.

References

B. S. Bloom (ed.), (1956) *Taxonomy of Educational Objectives*, Longmans.

D. Falk (1971) *Biology Teaching Methods*, John Wiley.

R. Mager (1962) *Preparing Instructional Objectives*, Fearon.

Further Reading

'Biological Sciences Curriculum Study' (1970) *Biology Teachers' Handbook*, John Wiley.

R. M. Gagné (1965) *The Conditions of Learning*, Holt, Rinehart & Winston.

D. R. Kratwohl, B. S. Bloom and B. B. Masia (1964) *A Taxonomy of Educational Objectives, Handbook II, Affective Domain*, David McKay Co.

H. H. Walbesser (1963) 'Curriculum Evaluation by Means of Behavioural Objectives', *Journal of Research in Science Teaching, 1, Issue 4.*

Independent learning

DONALD REID and PHILIP BOOTH

The alternatives to class teaching

'Pull the board down a bit please, sir—I haven't finished that top bit yet'.
'No—leave it where it is—I'm on the bottom line'.

'Now, as I was saying—Jones, stop interrupting, I'm not going to warn you again'.

Class teaching—we all know its drawbacks. But how far should it be replaced, and what with? The great bulk of all school teaching is class teaching—yet we hear of whole schools which claim to practise independent learning in every subject. Is this a feasible proposition for the ordinary overworked biology teacher in a typical secondary school?

The first step is to define the term independent learning. Independent learning takes place whenever a group of pupils is working at its own pace; during class teaching, on the other hand, all pupils are likely to be carrying out the same activity at the same time.

We prefer the term 'independent learning' to 'individual learning', since 'individual learning' seems to be associated with the idea of silent figures working in monkish seclusion. In practice, most pupils work naturally in pairs or small groups when working at their own pace.

Short term independent learning—within the duration of one lesson—is now common in many British schools, owing to the frequent use of worksheets to provide step by step instructions for practicals. Medium term independent learning, over a span of two to five double lessons (two to three weeks' work), is gradually gaining ground, but long term independent learning (four weeks or longer) is very rare.

Short term independent learning (one lesson)
Many lessons in biology and elementary science follow this
pattern:

> Introduction by teacher.
> Practical, following instructions on worksheet.
> Written work, perhaps based on questions from the
> worksheet.

Worksheets, in their simplest form, are simply a means of
preventing the waste of time involved in writing instructions
on the blackboard at the start of each lesson.

You can now use your microscope to have a closer look at
yeast.

(a) Collect a glass slide, cover slip, and mounted needle.

(b) Make sure your microscope lenses, and your slide and
cover slip are clean and dry — always check this whenever you use a microscope.

(c) Collect a small drop of yeast culture on your needle,
and put it on your slide.

(d) Place a cover slip on the drop, by lowering it carefully
with your needle.

(e) Look at the cells, first with low power, and then with
high power. Don't forget to let in more light when you
change over.

(f) If the cells can be seen clearly, *draw two or three of
them.*

Figure 1

However, use of even the simplest worksheets brings great
economies in staff time and the opportunity to raise
standards throughout a department. For example, a teacher
may be able to use the same sheet with different classes; or,
better still, his colleagues may use his worksheets with their
classes.

In many comprehensive schools, worksheet writing teams have been set up to provide material for mixed ability work with general science courses in the first three years. Each team member is required to produce high quality materials for a selected number of topics, but relies on his colleagues' efforts when teaching other topics.

If possible, the team holds regular meetings (e.g. Wreake Valley College, Leics.) to criticise the work in hand and to seek to improve the overall quality. Cooperative effort along these lines soon results in the creation of an extensive file of good quality material, which can be used immediately by even the rawest recruit to the department.

Where worksheets are being prepared for large-scale use in this way, the time saved can be employed to raise standards of presentation. Colleagues or ancillaries with artistic ability can be asked to provide illustrations, resulting in high quality materials such as these on the following pages Specialists teaching their own subject are less likely to form tightly organised teams, since they often prefer to produce their own materials. Nevertheless, we believe that even specialists should be encouraged to file all their work centrally and to hold periodic exhibitions of their work-sheets. In our experience, even those who prefer to compose their own materials always benefit by first looking at other people's work.

If worksheets are going to be anything more than a substitute for a few hasty instructions scribbled on the board, they need careful planning and meticulous presentation. For those who lack experience in this field, we list below our guide to the preparation of worksheets:

1 Decide on your objectives. Are you trying to achieve too much with the class concerned?

2 Write out the whole worksheet in rough first. Do not omit this vital step—it always saves time in the end.

3 If you are including questions, make sure that your questions fit the objectives.
  Check for ambiguities—try them on a colleague first, or better still, on two or three pupils, if you have the opportunity.

4 Check the instructions for length. Less able groups

EXPERIMENTAL WORK 2

Using potometers.

POTOMETERS are apparatus designed to determine the amount of water taken up or <u>absorbed</u> by a plant.

The <u>weight potometer</u> can also be used to measure the amount of <u>water lost</u> by a plant.

A. Using the weight potometer.

1 Set up this apparatus —

2 <u>Weigh</u>, and <u>record</u> its weight.

3 <u>Mark</u> the water level in the conical flask.

small plant
with well washed roots

layer of oil

conical flask

water

a) What conclusions would you come to —
— if, after about a week,
   i) the weight of the apparatus <u>decreased</u> by 10 grammes.
   ii) 11 cubic centimetres of water needed to be <u>added</u> to the conical flask to bring the water level back to its original marked level.

4 Leave the apparatus on the side bench, or in a green house for about one week.
   Meanwhile —

5 Draw up a results table to record —
   • the original weight of the potometer • the weight after one week
   • change in weight during one week • volume of water required to to restore the original volume.
   From your final results —
b) How much water is absorbed by your plant?
c) How much water is lost by your plant?
d) How do your results compare to others in your class?

simply won't read long, multi-step worksheets, so omit any less important pieces of information with these groups.

5 Always type the final draft if possible. If not, then write it out in a simple and legible style. Adult handwriting is often incomprehensible even to adults.

6 Check carefully for errors. Any slight mistake can cause endless problems with weak readers. ('Is there meant to be a word here, sir?')
Spirit masters can easily be corrected with a razor blade.

7 Illustrate as much as possible, especially if the worksheet is intended for use with the less able.

8 The method of reproduction needs careful thought. Spirit duplication (e.g. Banda) is simple and quick, but provides a limited number of copies and results in a worn-out master.

For top quality treatment, we recommend you to write out your master copy in Indian ink on ruled paper. A Roneotherm, or similar stencil, can then be prepared from it by passing the master through a heat copying machine, which will not reproduce the lines on the paper.

Use of this method results in a permanent master which can be easily amended by the use of white gummed labels. For high quality worksheet preparation (e.g. fig. 2 on p.156) we can strongly recommend this technique.

Medium term independent learning (two to five lessons)
The use of independent learning within one seventy-minute lesson is becoming common in science teaching because it does not call for any fundamental change in the role of the teacher. He is still responsible for generating interest in the work at the start, and for organising a discussion or summary at the end. During the short span of time involved, the fastest and the slowest workers in the class will not have become unduly separated, so that the teacher still feels fully in control of the situation. However, when independent learning is used over a longer span of time, e.g. two to five seventy-minute lessons, or about two weeks' work, there is a fundamental change in his relationship to the class. Both the initial motivation and any end-of-topic summary may be

widely separated in time. In addition, the fastest and the slowest workers will become widely spaced out, and the teacher may begin to lose confidence in his control of the situation.

To overcome the problems of motivation, adequate summaries, etc., very well prepared material is required. Certainly good quality illustrations will be very desirable, unless the work is exceptionally interesting in itself. To prepare material of this standard is always time-consuming, whether the illustrations are drawn by the teacher or pirated from a text book.

Consequently, there are considerable obstacles to overcome before independent learning can be confidently introduced over a span of two or more lessons. Nevertheless, we believe that there are strong reasons for making the effort to remove these difficulties, at least for the teaching of certain topics. Our confidence in 'medium term' independent learning stems from our experiences between 1968 and 1971, when we organised the Independent Learning Project in Biology, supported by grants from the Nuffield Foundation's Resources for Learning Project, Heinemann Educational Books Ltd, and International Computers Ltd. The purpose of the project was to write materials for medium term independent learning in eight selected topics. The eight draft texts were then tested in eighty British schools.

The conclusions of this project (Reid and Booth 1971) were that medium term independent learning is a valuable approach for certain types of topics, in particular:

1 topics involving the acquisition of a skill (e.g. how to use and make a key).
2 structured topics (e.g. surface area/mass ratio and its implications).
3 non-experimental topics (e.g. human reproduction or defence against disease).

Results from the eighty trials teachers, and from two controlled experiments (Reid and Booth 1971), suggest that independent learning over several consecutive lessons is preferable to class teaching for the topics mentioned, because:

1 Able pupils work faster than usual.
2 Learning and understanding may be improved in some cases.
3 Under the conditions of independent learning, teachers find that their knowledge of individual pupils is greatly improved.

Independent learning can also be recommended for several other obvious reasons; for example, it is a very suitable approach for mixed-ability classes when tackling difficult concepts. It is also invaluable where class teaching is difficulty to carry out, a situation which is by no means infrequent. We have come across many badly disturbed adolescents who work well during independent learning, but who habitually disrupt almost any class discussion in which they participate. Independent learning in biology is therefore especially welcomed by inexperienced teachers and also by non-specialists, e.g. a probationary psysicist nervously approaching his first biology topics in a Combined Science course.

However, we ourselves favour the use of independent learning neither for intellectual reasons nor for its tactical advantages. We use it because both we and our pupils enjoy it. This enjoyment arises chiefly because it represents a change from class teaching; improves our relationships with the class, and often allows hard working introverts to obtain the attention which they are denied by the more extrovert during class teaching. As an additional method of teaching, to be set alongside and used in conjunction with other approaches, we highly recommend independent learning over a two to five-lesson span.

Nevertheless, successful implementation of independent learning requires some degree of initial organisation. As we have mentioned, the first problem lies in the preparation of suitable materials. We favour a mixture of well-illustrated worksheets and programmed sequences, and armed with ample funds and time, we have been able to publish our own efforts. The example on page 160 illustrates the programmed approach. A typical example of an illustrated worksheet can be found in Book 3 of the same series (see page 161).

Collect 16 small cubes—

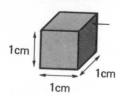

The surface area of each side is 1 square centimetre.

1cm
1cm
1cm

What is the surface area of the whole cube?

6 square centimetres (there are six sides)

Now join two cubes together—

Imagine you are looking at a model of an animal—

What is the area of its skin?

10 square centimetres (if you're not sure why, ask your teacher straight away).

1 Now use any number of cubes up to 16, to make a model of a tall, thin man—

2 Find the surface are of this shape and **write it down**

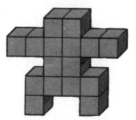

3 Now use **exactly the same number of cubes** to make a model of a short, fat man—

4 Find the surface area of the fat shape and **write it down.**

Both models have the same mass but which has the greater surface area?

Why do you think a tall, thin man may feel the cold more quickly than a short, rounded man?

First you will need to dissect a fish—

Read Instructions 1 — 4 before you start.

1 Collect a sprat (or other small fish),
a scalpel or sharp knife, and a dissecting board.

2 Lay your fish on its side, and cut through from **one side**
to the **other side**, as shown here—

3 Remove the top half of the body—

4 Your fish should now look like this—

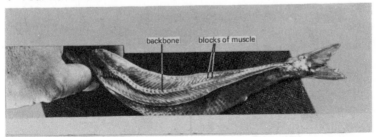

backbone   blocks of muscle

**Make a drawing of your fish to show the blocks of muscle
and the backbone.**

The production of home-made material to this standard is not easy, but it should be possible to tackle one or two key topics each year, even with a full time teaching load. Since our project ended, we have managed to prepare several topics for medium term independent learning with fourth and fifth years, despite the restrictions of full timetables. Key topics tackled in this way include:

Photosynthesis
Water relations
Feeding methods
Animal Nutrition

We hope to publish some of these titles in due course.

The construction of programmed materials is especially time-consuming. We would refer anyone interested in this technique to our advice on p.34 of the 'Introduction to the Series', *Biology for the Individual* (Reid and Booth, 1971). Despite the difficulties, we strongly recommend making the attempt to produce suitable material for independent learning in one or two topics. The use of good quality, home-made materials always brings a special sense of satisfaction to both teacher and pupils. Your prestige will be increased as a result; and your own material will always be better suited to the needs of your own pupils than the best published texts.

We recommend you to start with a topic which has a high theoretical content, but still contains some simple practical work likely to produce objective results, for example, osmosis. We do not recommend topics involving long-term experiments (e.g. growth studies), or topics where large scale results are required for valid conclusions (e.g. microbiology). Our texts on microbiology are primarily intended for short-term independent learning only.

Once suitable materials, whether purchased or home-made, have been obtained, it is necessary to give some thought to laboratory organisation. It is obviously important to follow these rules:

1 Try to assemble all equipment likely to be needed for a given spell of independent learning before the first lesson.
2 Set out the equipment in the same place in the room at

the start of each lesson, so that the pupils soon know where to look for it.

3 Train the class ruthlessly to leave stock items, such as sellotape or solutions, where they find them, and not to carry them round the laboratory.

4 Try to keep the items assembled as a kit between lessons, as far as possible. Trays and trolleys will be found very useful for this purpose.

5 Do not allow film loops to be viewed independently, unless yours is a very rich school. Repeated individual viewing leads to excessive wear. In any case, most film loops require a commentary.

Finally, considerable thought needs to be given to the role of the teacher during independent learning. It is, of course, essential for the teacher to participate fully in the work, if the advantages of this approach are to be obtained. Teachers who look forward to a spell of independent learning as an opportunity to catch up with marking, will soon find themselves dealing with some very bored and frustrated pupils. It is also a mistake to allow the written instructions to dominate the proceedings.

Our advice on the role of the teacher during independent learning is as follows:

1 The teacher must at all times retain a feeling of responsibility for the work and be ready to intervene whenever the materials seem unsuited to the needs of the pupils.

2 The children must be carefully initiated into this method of working; it should be stressed that they are now primarily responsible for their own progress.

3 The teacher should constantly circulate to discuss progress, stimulate critical thinking and chivvy the less enthusiastic.

4 The teacher should take every opportunity to vary the approach, even during spells of medium term independent learning. This can be done by showing film loops, or holding class discussions at convenient intervals.

Disadvantages of medium term independent learning

The major drawback to the use of independent learning lies in the attitude of the less able (e.g. the lower third of a comprehensive school's ability range) and the poorly motivated, of any ability.

The less able will include many who do not enjoy reading, and who will not always follow instructions carefully. They will either skip sections altogether or become bogged down in the first stages of the work. It is important to keep a close watch on less able pupils working independently in mixed ability groups; sometimes it may be desirable to permit them to omit the more theoretical parts of a topic.

Alternatively, it may be possible to divert them to simpler work on a similar theme. For example, the less able will probably prefer to find out about the habits of Arctic animals while their abler classmates are studying the importance of surface-area/mass ratio for Polar animals. However, diversion of the less able in this way needs to be carried out tactfully, by suggestion rather than compulsion. Those who wish to tackle the more difficult work should be allowed to attempt it, or the purpose of mixed ability work will be lost.

Very weak readers will benefit from good illustrations when following worksheets. Nevertheless, it may be necessary to provide tape-recordings of theoretical sections for use by slow readers—this is often a popular step.

Badly motivated pupils present more of a problem. They also need to be regularly chivvied, and it may be useful to set 'targets' to be reached by them during a given lesson. It is also true that in any given class, while some pupils actively prefer independent learning, others will prefer class teaching for the same topic. The proportions vary from topic to topic—a pupil who dislikes independent learning for plant reproduction, may prefer this method when studying human reproduction. Research suggests that introverts tend to prefer independent work while extroverts prefer class teaching; but there are many exceptions to this.

The obvious conclusion for the class teacher is to vary the work as much as possible. Consequently, independent learning and class teaching should be regarded as alternatives to be used whenever appropriate. Short term independent learning, based on a single period, will probably always be the commoner method, since this allows for class teaching within the same lesson. However, we believe that there is a strong case for spells of medium term independent learning, primarily as a means of introducing variety into the work.

Long term independent learning

We define long term independent learning as a situation in which class teaching occurs only rarely, and where independent learning is the dominant method. During long term independent learning, the pupils will be working independently for at least four weeks at a time, and may even be asked to work continuously for a year at a time. It will be gathered from our previous comments that we are not in favour of this approach.

Nevertheless, several projects have been set up to investigate the value of long term independent learning in biology and related subjects. We believe that the quest for success with this approach will continue to attract explorers in the future, and it is therefore worth discussing the advantages and disadvantages of long term independent learning, however unlikely an idea it may seem to the average biology teacher.

Long term independent learning was a goal which attracted much support during the nineteen-sixties. With the coming of the computer, the invention of the heady term, 'educational technology' and the rise of non-streaming, Utopia seemed just around the corner. One could visualise a budding genius taking O-level biology at twelve, A-level at thirteen, and perhaps a degree at 16—all without leaving his unstreamed class.

We ourselves shared in these dreams when we conducted an experiment in long term independent learning during 1967—8 for the Nuffield Foundation's Resources for Learning Project (Reid and Booth, 1969). At the same time, the Ford Foundation in the USA provided a grant, through its 'Experimental Program in Teacher Education', for experiments in the individualisation of some of the Biological Sciences Curriculum Study Group (BSCS) materials at the University of Colorado. The BSCS syllabuses and texts represent a development parallel to the Nuffield Biology Scheme.

Both the American experiment and our own project came, independently, to the same conclusion: 'We've seen the future and it doesn't work'. The chief drawback of long term independent learning in any subject is obvious enough—exclusive use of this method by itself leads to boredom. Certainly our less able pupils became more and more frustrated as they ploughed their way through endless sheets of paper, whether worksheet or programme.

In addition, there are a number of drawbacks peculiar to biology which make it an unlikely subject for the successful application of long term independent learning. The most obvious difficulty lies in the provision of an adequate supply of livestock. For example, when a class is about to study locust life-cycles as a group, it is sufficient to order one batch of hoppers, whose development to adults can readily be observed by the class as a whole. However, when the class have been working independently for several months, the ablest may reach the work on life-cycles many weeks ahead of the least able. As a result, it will be necessary to maintain stocks of locusts at all stages of development until the last person has completed this section of the work. Since the same problem will be encountered with frogs, rats for dissection, plants, etc., much extra work will be entailed for both teachers and technicians.

In addition, many biological experiments need to be repeated on a large scale before valid conclusions can be drawn (e.g. experiments in microbiology). This presents no problem to a class working together, but individuals cannot obtain valid results without costly and time-consuming repetition of their experiments (e.g. using twenty agar plates instead of two). Consequently, long term independent learning in biology is extremely difficult to organise, calls for expensive duplication of resources and is of doubtful validity. Since much of the work in elementary biology is less conceptual than in the other sciences, we believe that there is no place for long term independent learning in biology at school level.

Only two successful examples of long term independent learning in related fields are known to us. In mathematics, where none of the practical problems apply, the Kent Mathematics Scheme to provide independent learning from eleven to sixteen is steadily gaining in popularity in schools far from Kent. In the USA, the Intermediate Science Curriculum Study (ISCS) based at the University of Florida (ISCS 1972) has produced a complete independent learning course in general science for the eleven to fourteen age range, roughly equivalent to the age and syllabus content of Nuffield Combined Science. The ISCS scheme appears popular with American schools which contain a good proportion of abler pupils. However, much of their well-produced material is concerned with physical science or with elementary geology.

We therefore believe that long term independent learning is inappropriate to the teaching of biology in schools. The difficulties involved in its organisation outweigh the advantages. However, the opposite may be true in physics and mathematics, provided well-designed materials are used, and attention is given to the problem of overcoming the inherent lack of variety in such methods.

SUMMARY AND CONCLUSIONS

1 Both class teaching and independent learning have a place in the teaching of biology, though neither should be used to the exclusion of the other. We believe that variety of method is always of great importance.

2 Short term independent learning, within the span of one seventy-minute lesson, is easily carried out by means of worksheets. If possible, these should be prepared on a team basis.

3 Medium term independent learning, over two to five lessons, presents a greater contrast to class teaching, and results in faster working for abler pupils and a closer relationship between teacher and pupils. However, it requires care in preparation.

4 Long term independent learning, over four weeks or longer, is not recommended in the teaching of biology, owing to the difficulties of providing adequate living material and the need to conduct many experiments on a large scale. It also leads to a lack of variety, in the long run.

References

D. J. Reid and P. Booth (1969) 'Using individual learning with the Nuffield Science Course', *School Science Review*, vol.50, no.172.

D. J. Reid and P. Booth (1971) 'Independent learning in Biology—results of trials, 1968—70', *School Science Review* vol.52, no.180.

D. J. Reid and P. Booth (1971) *Introduction to the Series, Biology for the Individual*, Heinemann Educational Books.

D. J. Reid and P. Booth (1972) *Biology for the Individual, Book 5, Problems of Life in hot and cold climates*, Heinemann Educational Books.

Intermediate Science Curriculum Study (1972) Silver Burdett & Co.. Morristown, New Jersey, USA. (A wide selection of texts and teacher's guides.)

Quantitative work
and problem solving

BRIAN DUDLEY

Introduction: biology and mathematics

In secondary schools biology is taught essentially as a
descriptive subject, even though it is classed as a science. The
reasons for it being descriptive are to be found in its origins
and in its development within the school curriculum, where it
became established largely as that science suitable for pupils
who had been declared unable to cope with mathematics—
and by implication for only such pupils—since those profes-
sing an ability in mathematics were guided into the other
sciences, physics and chemistry.

Biology and mathematics developed quite independently
as subjects within the education system, at least in Great
Britain, and this, until recently, seems to have been viewed
with some satisfaction by both parties. Indeed, so separate
have they become that an almost fond belief exists that
ability in the one field precludes that in the other.

Under such circumstances it was inevitable that biology
came to be timetabled simultaneously with mathematics at
the sixth form level, so setting the seal on this fond belief by
precluding their common study. Then the system became
largely self-justifying, with some of these mathematically less
inclined pupils becoming biology school-teachers themselves,
and perpetuating the non-mathematical aspects of the subject
to the oncoming generation of similarly selected pupils. In
short, there has grown up within schools a low expectation of
performance in mathematics amongst pupils studying biology
and this remained unchallenged for decades, and apparently
became more unassailable with the passage of time.

At first sight their mutual isolation appears to be soundly
based and valid. For instance, in contemporary theories of
the structure of knowledge, mathematics and science are
placed in different categories and hence represent fundamentally different ways of acquiring knowledge. From a study of

the structure of abilities it seems that science (biology) requires verbal reasoning, inductive and divergent thinking, whilst mathematics requires non-verbal reasoning, deductive and convergent thinking. Both in the way of acquiring knowledge and in the abilities required for it, mathematics and biology appear to be quite separate, and to be pursued by quite different types of person.

But mathematics belongs to that category of knowledge known as 'symbolic knowledge', the means by which all categories of knowledge are expressed. To this extent, mathematics is required to express scientific knowledge and, in consequence, a relationship between mathematics and biology must exist. Furthermore, Vernon (1961) shows that both scientific and mathematical abilities are composite and both are made up of essentially similar minor groupings in the hierarchy of ability factors which he has raised. Indeed, Vernon does not distinguish between abilities in mathematics and in science, but writes of a unitary mathematical-scientific ability. Evidently the two subjects are not to be isolated from one another on grounds of the inherent structure of knowledge, nor of the abilities required in their study. In fact, it is helpful to avoid thinking of people in terms of ability types; subjects make common use of a number of abilities, and both mathematics and science require abilities in order to solve problems which involve verbal and non-verbal abilities, as well as a variety of convergent and divergent thinking. There can be little doubt that both deductive and inductive thinking also are required.

Since there is no fundamental, structural reason why mathematics and biology should be isolated from one another in schools, the fact that they remain so today must, in large measure, be due to attitudes, and especially the attitudes of teachers. In 1965 a study showed that 59 per cent of teachers considered that the fact that biology was seen as a non-mathematical science was a frequent, or very frequent reason, for pupils taking biology in the sixth form. But, in 1972, a different study found that large numbers of students themselves maintained that this was not a reason for their taking up A-level biology (Kelly, 1972).

Between 1965 and 1972, there may have been some movement in the attitude of students, or perhaps, in 1965, the teachers were mistaken in their reading of students' attitudes. Certainly teachers were mistaken in 1971 during trials of the Nuffield A-level Biological Sciences Project, in their belief that students taking biology tended to have low mathematical ability, for this was shown by Kelly not to be the case. Also, teachers who believed that the mathematics included in the scheme was too difficult, as well as those who claimed that girls taking A-level biology were less able at mathematics than were boys, were shown to be holding convictions that had no foundation in fact. Later, the teachers came to the conclusion that, with regard to teaching the mathematics in the scheme, the problems encountered were not so much related to the level of the difficulty of the mathematics, or to the ability of the students, as they were to the learning strategies employed.

Perhaps it is the attitude of teachers themselves that is crucial in bringing mathematics into secondary school biology. Clearly, attempts to justify their total separation at this level are not easily sustained. However, there seems to be a need to encourage biology teachers by ensuring that any mathematics used in biology is in itself, at all times, both worth while and useful from their point of view, and also demonstrably within the comprehension of both themselves and their pupils.

Size and Shape

The need to bring mathematics into secondary school biology has been recognised for some time, and an attempt to do this was made in the Nuffield O-level Biology Curriculum Project (1966–7) where, in the second year, both mathematical and biological concepts were introduced in relation to size and shape. In spite of its desirability and validity on educational grounds, the actual work (as set out in the project) has created difficulties and adverse comment. Central to that work is the relationship between surface area and volume, itself expressed mathematically as a ratio. This ratio is continuously variable, representing as it does the different effects of size upon surface area on the one hand, and volume on the other. Conservation of volume is not always easily understood by children at this age. Furthermore, while volume is influenced mostly by size, surface area is influenced more by shape, so the ratio between the two is meaningful only for as long as shape is constant ('invariant', in mathematical parlance). Having set out to establish the effect of size upon surface area/volume ratios, the teaching scheme concentrates upon cross-sectional area (rather than surface area), and upon mass (rather than volume), between which there is a direct relationship only when density is constant (invariant). None of these factors is identified in the scheme and so it produces many difficulties of understanding. A more careful analysis of the concepts and relationships involved is helpful to teachers, but possibly more useful is a simpler conceptual framework for teaching the problems of size and form in biology. It is believed that one such simple framework has been found.

Scale enlargement

If, instead of the effects of increase in size upon the surface area/volume ratio, one concentrates upon the growth in size of organisms whose shape remains quite unchanged (invariant), then the study of size, a concept in biology, becomes at the same time a study of enlargement, a concept in mathematics. Enlargement is measured as the 'scale factor', a ratio, but one which is simple to comprehend since it is the ratio between corresponding linear measurements of the objects concerned, and also is invariant when they are identical in shape, whatever linear measurements are taken. Of course, in biology no two stages in the growth of an organism are *identical* in shape, nor are two similar organisms identical in this way. But if model animals are made which are scale enlargements of one another, then growth in size of living things can be taught from them.

Three models can be used, each with a metal container for the body, drinking straws for legs, and coloured, shaped paper for head, coat and tail (Dudley 1971). I have found it most effective to hand out the small one in class and ask for a list of what is needed to make a model twice the scale. When the lists are correct (heads are provided), the parts are given for pupils to assemble their own scale model. By repeating the exercise, making another of either scale factor x four of the original, or x two of the pupil-assembled model, all three come to be present in the classroom. It is then a simple matter to compare corresponding lengths—any lengths—and show that the ratio in every case is that of the scale factor. While a comparison of corresponding areas (coats are very suitable) shows that areas grow as the square of the scale factor. Volumes are found to grow as the cube of the scale factor. These relationships are easily drawn from the models.

But the containers are empty and this does not correspond to real life. Since much of the body weight of an animal is water, the containers represent animals when filled with water. Water-filled containers, one finds, cause the legs of the largest model—but not those of the other two—to collapse; the strength of the legs has been outstripped by the mass of the body as size increases. If terrestrial animals increase in size without changing shape (that is, if they grow by enlargement) then at some time they become mechanically unsound, because their legs do not increase in strength as fast as their body increases in mass. If the animal is to grow to this large size then some modification of its shape is necessary. How might this problem be investigated?

The load-bearing capacity of these legs can be studied by fitting one straw per leg to a model and testing to destruction with weights from the physics laboratory. Repetition with straws of different lengths soon gives enough information to be able to decide in what ways the legs of the large model might be modified to allow it to stand. It emerges that the legs, if they are to be of scale length must be extra thick, if of scale thickness then shorter than scale length, while they can be quite thin if they are really short. In the first case the model, when built, looks like an elephant; in the second, a rhinoceros; and in the third, a hippopotamus. The models are even more convincing when an appropriately shaped head is fitted.

Once the shape has been changed one has left enlargement behind, but the changes have been shown to be necessary and also realistic, since the large model takes on recognisable animal forms. The lion is said to be a 'big cat'. Is it a scale enlargement of a domestic cat? If it is then it has not had to modify its legs in order to support its increased mass (a cat weighs about twelve pounds and a lion about 400 pounds). In fact, the lion can be shown to approximate very closely to a scale enlargement of a cat and thus its description is shown to be apt. The fact that modification of shape becomes necessary only when the animals have become very large in size is also underlined.

That scale enlargement does occur in real life is illustrated further by two more examples (many others have been found):

1 A giraffe when born is 6 feet tall and weighs one hundredweight. An adult is 18 feet tall and weighs about 25cwt
Does this represent scale enlargement?
Scale enlargement of the adult giraffe = $x \dfrac{18}{6} = x\ 3$

Weight of new-born giraffe $= 1$ cwt
Assuming growth is by scale enlargement, the weight of the adult giraffe can be expected to be $(3)^3 \times 1$ cwt
$$= 3 \times 3 \times 3 \times 1$$
$$= 27 \text{ cwt}$$

This weight of 27 cwt compares favourably with the known weight of about 25 cwt and indicates that it is a reasonable approximation to claim that the growth of a giraffe is by enlargement. It certainly shows there is no marked change or modification of form during growth, and that the new-born giraffe is essentially similar in shape to that of the adult.

2 How much heavier is an African elephant than its Indian counterpart? A full grown Indian elephant stands 8 feet tall at the shoulder and weighs 3.5 tonne. A full grown African elephant stands 9 feet tall at the shoulder. How much difference does this extra 12 inches make to its weight?

Scale enlargement of the African elephant $= x \dfrac{9}{8}$

Expected weight of African elephant (tonne)
$$= \left(\frac{9}{8}\right)^3 \times 3.5$$
$$= \frac{9 \times 9 \times 9}{8 \times 8 \times 8} \times 3.5$$
$$= 5.0 \text{ tonne (to first decimal place)}$$

Because it is 12 inches taller at the shoulder, the African elephant is 1½ tonne heavier than the Indian elephant. At these sizes even small increases have marked effects upon the load which the legs have to bear.

So far the subject matter has been concerned with scale models, specially built to illustrate the effects of scale enlargement upon terrestrial animals, and with examples of existing animals whose shapes are such that it is sufficient to estimate that scale enlargement is likely to apply. But enlargement is an exact mathematical concept and can be tested for precisely. These tests too have their value in teaching biology and can lead to further mathematical concepts, more immediately recognisable to the biology teacher.

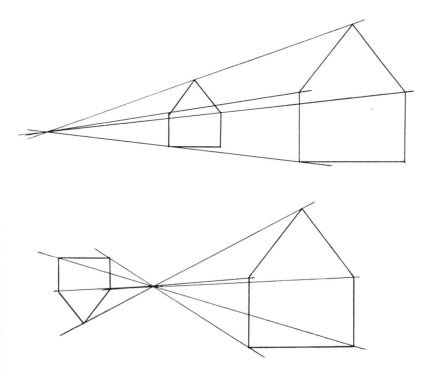

Figure 1 Testing for enlargement between shapes of different sizes

Testing for enlargement

If enlarged shapes are arranged side by side, and straight lines drawn joining corresponding points, then these lines will all converge and meet at a point, called the centre of enlargement. When these similar shapes are the same way up, this centre of enlargement lies to one side of them, but when one shape is inverted, it lies between them (fig. 1).

It follows that in any such test the formation of a centre of enlargement demonstrates that the materials tested are identical in shape, while any test in which a centre of enlargement fails to form demonstrates that the materials are not identical in shape. On this basis the growth of dicotyledonous leaves can be investigated once they have left the bud, and until they are full grown, to see if, during this period, their growth in size is by enlargement. While it will be found that a number of dicotyledonous leaves do not grow in this fashion, those of the laurel do (fig. 2).

Although a number of problems immediately arise as to how this growth might be achieved at the cellular level, and the consequences to the leaf's water balance should growth at that level be confined exclusively to the enlargement of existing cells, the point to be established here is that leaves which do grow by enlargement are ones for which leaf area is particularly easy to calculate.

Take one laurel leaf as a standard. Find its area by one of the usual empirical methods and measure the length of its midrib. The scale enlargement of any other laurel leaf will be given by the ratio between the lengths of the two midribs (corresponding linear measurements), while their areas grow by the square of the scale factor, as is true of the areas of any figures that have grown by enlargement.

If A = area of standard leaf (known)
  B = area of test leaf (unknown)
  C = length of midrib of standard leaf (known)
  D = length of midrib of test leaf (known)
Then $\dfrac{A}{B} = \dfrac{C^2}{D^2}$, when $B = \dfrac{A}{C^2} \times D^2$

But A and C are measurements of the standard leaf, So $\dfrac{A}{C^2}$ is constant (K).

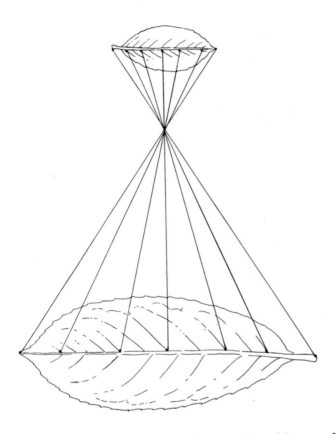

Figure 2 Testing for enlargement between laurel leaves of
different size.

Hence B, the area of the test leaf, is given by K x D² with D being the length of the midrib. The value of K depends upon the units of measurement adopted and on the exact shape under study.

Since the approximate areas of laurel leaves can be calculated without having to pick them or spend time measuring each area empirically, then the problem of the rate of transpiration from leaves under different experimental circumstances becomes a quantifiable study in the classroom, at least when laurel leaves are made the subject of the investigation.

Mapping

When testing for scale enlargement what is used, in fact, is a mapping—one point on one leaf, and its corresponding point on the other, are joined by a straight line. Mappings are used frequently in biology though teachers are more likely to recognise them as tables of results which represent the mapping of one numerical value in one column to a specific value (usually on the same line, or row) in an adjacent column, both values having been established by reference to some standard unit of measurement. This is a one-to-one mapping.

In a table with just two columns of results, if the numerical values in each column are first mapped onto a linear scale of the standard units involved, then these results can be presented in one of a number of diagrammatic forms. For instance the two scales can be arranged parallel, with the values increasing in the same direction. Straight lines can then be drawn between each pair of corresponding values and the direction indicated: one then has an 'Arrow' diagram. In another form the scales, while being parallel, can have their values increasing in opposite directions. In the first instance lines joining the related points will converge to one side of the scale lines (fig. 3a) while, in the second, they converge between the two scale lines (fig. 3b). In either case if the lines joining related points converge at a single point, then the numerical values of the two columns are in proportion. This means that ratios made up from the two values of each mapping are all equivalent.

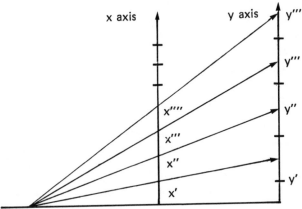

(a) A direct relationship

$$\frac{x'}{y'} = \frac{x''}{y''} = \frac{x'''}{y'''} = \frac{x''''}{y''''}$$

(b) A direct relationship with one scale inverted

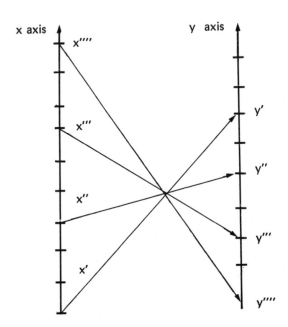

Figure 3 Arrow diagram involving numerical values which are in proportion.

Fig. 3b can also be used as a nomogram, a diagram for interconverting standard units (Fahrenheit/Centigrade; inches/centimetres; miles per hour/metres per second), one example of which is given in fig. 4. Any straight line drawn from either scale line through point 0 will meet the other scale line at its equivalent point.

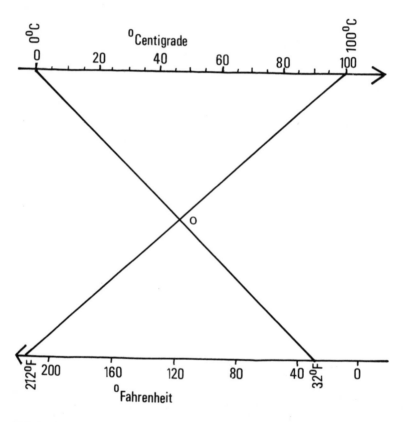

Figure 4 Nomogram for interconversion of Fahrenheit and Centigrade temperature scales.

Graphs

The same two scale lines (instead of being drawn parallel) can also be arranged at right angles to one another, with the lowest values of each at their intersection. If the mapping is then achieved by drawing straight lines at right angles to these scale lines, and their points of intersection marked, then we have the familiar two-dimensional (Cartesian) graph. Data, which on an Arrow diagram shows the lines all meeting at a point, will form on the Cartesian graph a series of points lying on a straight line (fig. 5).

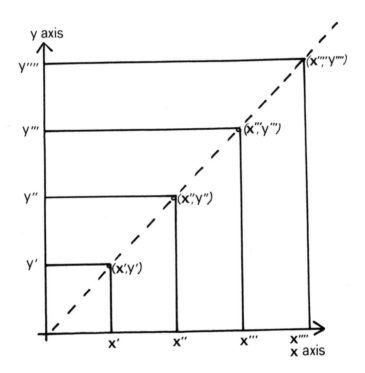

Figure 5 Cartesian graph

Straight line graphs are powerful tools in biology, being useful in the solving of many problems. This is because the equation of the straight line graph, $y = mx + c$, shows that, in addition to the two variables being measured in the experiment (x and y), there are two pertinent constants (m and c) and a very specific relationship exists between all four factors. Furthermore, the values of both constants can be identified from the graph itself; m is the slope, or gradient, of the line $\left(\dfrac{dy}{dx}\right)$ and c is its point of intersection with the vertical axis (y) (Fig. 6).

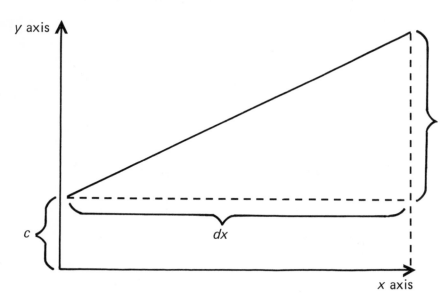

Figure 6 Analysis of the components of an equation giving a straight line graph.

In biological studies Cartesian graphs have curves rather than straight lines. However, if by changing the scale on the axes these curved lines can be converted into straight lines, then the above equation again applies, and the relationships between the variables and constants can be identified. The precise relationship depends upon exactly what is required to be done in order to get the graph linear. For instance, if this results from making the scale of both axes logarithmic (a log-log plot), the linear relationship is $\log y = m.\log x + \log c$. This, in another form, is $y = c.x^m$ and often is found in biological studies concerning surface area and volume relationships. In studies concerning only areas, m is 2; while in others concerning only volume, m is 3. But the effect of size upon the basal metabolic rate of mammals (y), for instance, is influenced by both area and volume simultaneously—the needs are related to the volume, and the supply of those needs to the surface area. If the surface area/volume ratio was the only factor involved, then m would be 2/3 (0.667). In fact m has been found to approximate more closely to ¾ (actually 0.72) and from this we know that additional factors must be involved. The nature of these additional factors remains to be discovered.

Should a linear graph be formed when the y axis is logarithmic and the x axis is linear (a semi-log plot), then the relationship is shown to be exponential, and the equation of that straight line is $\log y = x. \log m + \log c$. In another form this is $y = c. m^x$ (x is the exponent) and, once again, the two variables (x and y) are being measured and the two constants (m and c) may be identified from the graph. This relationship is often encountered in the growth of population. When c is the initial number of individuals in the population and growth occurs only by binary fission, m is 2. It is from the linear form of this relationship that the time taken for a generation to grow up and divide is established. Predictions of the future size of human populations are based upon these relationships and so are reliable only if their growth is, and remains, exponential.

Lattices

There is another way in which Cartesian geometry can help in the teaching of biology. In a Cartesian graph, any point is located by reference to two intersecting lines (coordinates), and its location is given by a pair of numbers (x,y). Hence any such point (p) is defined in terms of a set of two numbers (or elements), and can be written $p = \{x,y\}$.

In diploid cells any gene (h) consists of a set of two alleles and can be written $h = m,f$ where m is the paternally inherited allele of gene h, and f is the maternally inherited allele. So a gene can be represented by a Cartesian coordinate and a genetic cross, involving Mendelian inheritance, can be represented by a lattice (a graph-like figure) in which the types of possible offspring are identified by its coordinates.

For instance, a monohybrid cross between a pure

dominant parent (AA) and a pure recessive parent (aa) has this appearance:

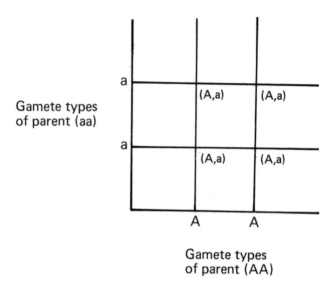

Gamete types
of parent (AA)

When both parents are heterozygous the genotypes of the offspring are:

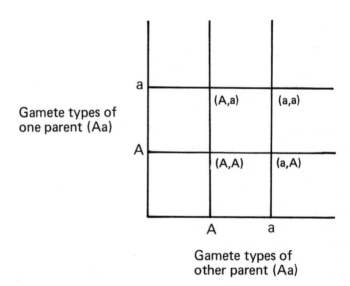

Gamete types of
other parent (Aa)

In this cross Mendel showed that the dominant and recessive phenotypes occur in the ratio of 3:1. This can be explained only by assuming that the gamete types of each parent are equally frequent. Bearing in mind that ½ x ½ = ¼, the lattice now becomes:

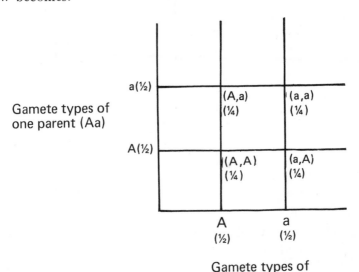

Classical mendelian ratios, both from monohybrid and dihybrid inheritance, can be dealt with in this way; so can special cases, such as multiple alleles, sex linkage and autosomal linkage (Dudley 1972), making possible an especially desirable cohesion for teaching Mendelian inheritance at the secondary level.

Population genetics also comes within its scope. The Hardy-Weinberg Law, in particular, can be taught as a single diagram, being concerned with the inheritance of a single gene (A) in a population of heterozygous parents. Because the whole population is interbreeding, the frequencies of alleles (A and a) amongst the gametes of each sex is unknown. Let them be p and q respectively. The effect of this on the frequencies of genotypes in the next generation is shown in the following lattice; and is found to be that stipulated by the Hardy-Weinberg Law.

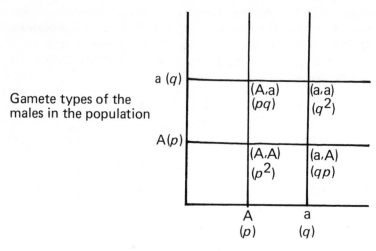

Gamete types of the
males in the population

Gamete types of the
females in the population

The Hardy-Weinberg Law is a special case of the binomial expansion $(p + q)^n$ in which n is 2. Mendelian inheritance (Mendel's Ratios) is a further special case in which $p = q = \frac{1}{2}$ and n is 2. The classical F_2 phenotype ratio, involving unlinked genes, is another special case in which p is 3, q is 1 and n is the number of genes in the system (Dudley 1973). If all the genes affect the same feature, we move from discontinuous variation to the continuous variation of a single feature. When such genes contribute independently and equally to one feature, the variation is symmetrical about a mean and shows a Normal distribution, another focal point in studies of both variation in biology and the statistics involved in its analysis.

Conclusions
Not one of these biological topics is new to the biology teacher; all are established as integral parts of the biology currently being taught in secondary schools. But each, traditionally, is taught in isolation. Links between them are not seen to exist. In fact, there is an underlying continuity of thought and this is essentially mathematical in nature. By restructuring these topics, and by making use of the

mathematics of them, the biology teacher has the opportunity to augment the coherence of what is taught. While the content of courses would not be affected, the mathematics would help to identify, and then solve, what are, without question, biological problems.

References

B. A. C. Dudley (1971) 'Teaching size and form in biology', *Journal of Biological Education*, 5.

B. A. C. Dudley (1972) 'The Mathematical basis of mendelian genetics', *Journal of Biological Education*, 6.

B. A. C. Dudley (1973) 'The mathematical basis of phenotype ratios', *Journal of Biological Education*, 7 (in press).

P. J. Kelly (1972) 'Evaluation Studies of the Nuffield A-level Biology trials, 3: Student characteristics and achievement', *Journal of Biological Education*, 6.

P. E. Vernon (1961) *The structure of human abilities*, 2nd edition, Methuen.

Radioisotopes in biology teaching

DAVID HORNSEY

Introduction to the properties of Radioisotopes

The intention of this chapter is to show, and perhaps convince, the biology teacher that there may be a place in class work for biological investigation employing radio-isotopes. It is important to realise in this context that the isotope is a tool and with it the biologist has been able to make observations that would have been impossible without it. In some cases this tool has required complex and expensive apparatus to ensure its detection. In other cases, detection may be made without resort to electronic appara-tus at all (for example autoradiography). Between these two extremes lie the simple radiation detectors that are an economic possibility in school work and which allow a considerable increase in the sensitivity of experimental work. Before going into the details of the merits of radioisotope usage in experimental work, it may well be useful to summarise briefly some important physical properties of the radioisotope.

Radioisotopes are unstable elements which regain stability by the release of energy in the form of *radiation*. There are three basic forms of this radiation: *alpha-, beta-,* and *gamma-*radiation. In the main the latter two are the ones used by the biologist. The beta-particle is a high speed electron emitted from the nucleus. The gamma-photon is a release of electromagnetic radiation, due to a rearrangement of the energy state of the nucleus. Gamma-radiation is very pene-trating in both air and solid materials, whereas beta-radiation may be easily stopped in materials such as perspex, glass, aluminium, etc.

The radioactive elements of major importance in biology are hydrogen-3 (tritium), carbon-14, phosphorus-32 and sulphur-35. All are pure beta-emitters with no associated gamma-flux. The use of such isotopes is an immediate advantage in laboratory practice because one is dealing with a weakly penetrating form of radiation.

An isotope which decays to stability may have a limited shelf-life. Such a concept is expressed in terms of the *half-life*. This is the time taken for a large number of radioactive atoms to decay to half their original number. The half-lives of some of the common isotopes are given in Table I.

Table 1 Half-lives of some common radioisotopes used in biology.

Hydrogen-3 (tritium)	12.3 years
Carbon-14	5760 years
Sodium-22	2.6 years
Phosphorus-32	14 days
Sulphur-35	87 days
Iodine-131	8 days

Because of their biological importance, their relative ease of detection and the fact that they behave chemically in the same way as their stable counterparts, these isotopes have proved invaluable in solving numerous physiological and biochemical problems. The carbon pathway in photosynthesis is a classic example, but investigations into molecular biology, metabolic studies and ion transport across membranes have all owed their solution in some way to the radioactive isotope.

The activity of radioisotopes
The major reason why the radioisotope has proved of such value is the sensitivity of its detection. When purchased from a supplier, several pieces of information are supplied with a particular isotope. Firstly there is a measure of the *level of activity*. The units of this activity are *Curies (Ci)* or sub-multiples, *millicuries* or *microcuries (mCi or μ Ci)*.

1 Curie is equivalent to 3.7×10^{10} nuclear disintegrations occurring per second (dps).

1 millicurie is 3.7 x 10^7 dps
and
1 microcurie is 3.7 x 10^4 dps or 2.2 x 10^6 disintegrations
per minute

The curie is a large level of activity and is seldom used in biological studies. The millicurie (still a relatively large level) is usually the amount maintained as 'stock'. The microcurie is the amount of the isotope that is added to the biological system under investigation. 1 μCi of (say) phosphorus-32 is about as much activity as a detector, such as a Geiger-Müller counter, can successfully measure. It is important that in any 'tracer' studies only small levels of active materials are used.

The second piece of information that is relevant in the case of isotopes of limited life is their *date of preparation*. This information allows corrections to be made on the level of activity left after storage. The decay of an isotope is given by the expression: $A_t = A_o e^{-\lambda t}$
where A_t is the activity left after a time period t, A_o is the original level of activity and λ is a constant known as the decay constant.

Corrections using this decay law often cause the biologist considerable anxiety. This may be remedied by a simple plot of the logarithm of the percentage decay against the time of decay. Because phosphorus-32 is the isotope that is most likely to cause difficulty, such a plot appears in fig. 1. As an example, 50 μCi of ^{32}P was purchased on 2 October to be used on 8 November. Thirty-seven days have elapsed since its preparation and, from the graph, only 16 per cent of the original activity remains; that is 8 μCi.

The final piece of information supplied with the isotope is known as the *specific activity* and it is this that represents one of the basic reasons for the use of radioisotopes in experimental work where great sensitivity is required. The specific activity of an isotope is the amount of activity expressed per unit weight of the element or compound. Again if we take ^{32}P as the example, the specific activity may be of the order of 60 Ci of ^{32}P per mg. of phosphorus. This means that for every mg. of phosphorus present there would be 60 Ci of activity. This quantity will not change with time, as will the activity, because when ^{32}P decays it

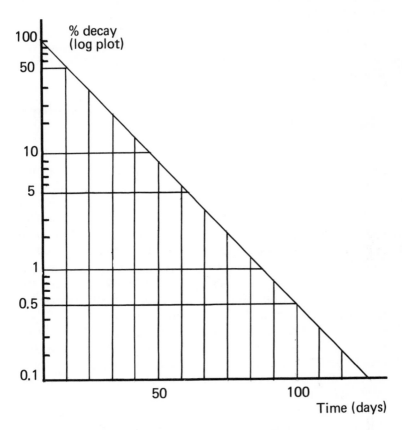

Figure 1 Decay curve for phosphorus-32

produces the new element ^{32}S; therefore the ratio of activity to weight of phosphorus will remain constant.

Let us examine the implications of using such material. If 1 μCi of ^{32}P is placed near a Geiger-Müller counter, then the activity recorded may be of the order of 480,000 pulses or counts per minute. As pointed out previously, this is a little too much activity for the detector but will serve as an example. 1 μCi of the quoted specific activity will have a weight of 1.6×10^{-8} mg. of phosphorus. The detector is quite capable of detecting levels of activity down to 0.001 μCi, and this is equivalent to measuring 1.6 x 10^{-11} mg. of phosphorus. With some fairly simple and relatively inexpensive counting apparatus then great sensitivity can be introduced into the school laboratory. Chemical methods of analysis to achieve anywhere near this sensitivity would be prohibitively expensive.

Here then is a major argument for the use of radioisotopes. What are the others?

Radioisotopes in schools

Students who read articles, like those appearing in the *Scientific American, New Scientist* and other such periodicals, must constantly be aware of the involvement of radioactive tracers in fundamental research. Such references will probably prompt questions about these tools of research and hence their introduction into the school curriculum would be an educational benefit. As part of his curriculum the physics student becomes involved in nuclear physics and the physical properties of radiations and radioactive materials. Practical work in this study is mandatory. This fact is of importance to the biologist, because not only will he have some assistance from his physics colleagues on any problems that may arise in the introduction of the work, but also, in all probability, a radiation detector will be available in the school for his use.

The immediate objection raised against the use of radioisotopes, and particularly their introduction into schools, is the hazardous nature of the radioactivity. This is not the place to discuss in detail the complex subject of radiological protection. Radioactivity, as with any other agent—electricity, gas, fire—may be a hazard, but with an understanding of the few precautions needed in handling, any hazards may be reduced to almost negligible proportions. In any biological experiment one is dealing with an unsealed or open source of radioactivity. This means that it is in the form of a gas, powder or more usually, a liquid. The implication of this is that such a source may contaminate apparatus and the person. Hence the need for special knowledge of handling. The hazard from such sources does not usually arise from the exposure of the body to external radiation, mainly because of the weak penetration of the radiations involved. It comes from the possible ingestion of the radioactivity via skin contamination. If such materials are treated as poisons—no mouth operations, particularly pipetting, the wearing of protective gloves and a good standard of cleanliness—then any hazard will be considerably reduced. Although I have used the poison analogy with reference to the handling of radioisotopes, this is as far as the similarity goes with respect to the effects of these materials on the body. The ingestion of poison without rapid treatment may well be fatal, and is certainly immediate, whereas the ingestion of small amounts of a radioisotope is neither fatal nor immediate in its action. In this connection, I will refer the reader to the medical uses of radioisotopes in diagnostic investigations (Hornsey, 1973). For further information on radiological protection and its practice in the school laboratory, reference may be made to the basic text (Andrews and Hornsey, 1972).

Another of the reasons for a reluctance to introduce experiments with radioisotopes, and one not often voiced, is a lack of confidence in such an apparently alien field of study—alien certainly to many biologists. To overcome this problem, several short courses are run at Harwell and at some universities (such as Bath). These courses, usually a week long, cover the basic principles of radioisotope work for teachers with no previous experience in the field. At the end of such periods of instruction, the biologist, unlike the chemist or physicist, finds that the use of isotopes can open up new teaching concepts. Kinetic experiments become a relatively simple exercise and with this the dynamic nature of biological systems becomes even more apparent. I find this particularly so when making observations on the movement of ions across living membranes. The placing of radioactive sodium (Na-22) in the water surrounding a membrane (frog skin) and the sampling and assay of the radioactivity, as it appears in an isotonic salt solution on the other side of the membrane, allows an investigation into the active transport of sodium ions against a considerable concentration gradient. The use of inhibitors and metabolic poisons on this 'uphill' movement means a greater understanding of the mechanisms involved in such transport, and also confirms the dynamic state of living organisms. Such investigations can form the basis of a study into the ionic regulation of animals living in estuarine and fresh water. This kind of project is realistic in schools near aquatic environments but is not so practicable for those less favourably situated. Under these circumstances physiological investigations into photosynthesis, respiration and ion movements across the membranes of plant roots are very valuable alternatives.

Investigations involving plants

Experiments with plants are an obvious attraction because of the very nature of the material. No Home Office licences are required, as they are with vertebrate animals. As a zoologist, it was not until I conducted physiological experiments with whole and isolated plant systems, that I discovered how dynamic the sedentary plant was.

One of the classic experiments in biological investigation is the tracing of the carbon atom through the photosynthetic pathway. In a simple way, these experiments can form the basis of a series of school experiments. Plants are first exposed to an atmosphere of carbon-14 labelled carbon dioxide for approximately two hours. At the end of that time the leaves are examined for the presence of radioactivity. Before such an experiment is described in more detail, it may be useful to anticipate the question—is it possible to get the same results without recourse to radioactivity? A non-radioactive experiment, such as the starch-iodide test on leaves exposed to light and carbon dioxide, will indicate the need for such agents in the process, but the carbon-14 experiment will show the actual involvement of inorganic carbon in the manufacture of organic compounds. I find this scientifically better than obtaining the results by inference.

This experiment has been reported several times (Andrews and Hornsey, 1967, Paice, 1968, Nuffield, 1970). Leaves of plants, variegated or not, are confined to a bell-jar in which carbon-14 labelled carbon dioxide is generated by the action of acid on carbon-14 labelled sodium carbonate (fig. 2). During the experiment the bell-jar should be kept either in a fume-cupboard, or near an open window of a well-ventilated laboratory, or outside. The levels of activity used will depend on the extent of the investigation. 1 μCi of $Na_2{}^{14}CO_3$ will only allow studies into the distribution of activity in the leaf

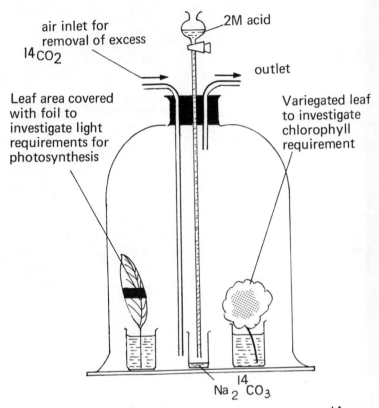

air inlet for
removal of excess
$^{14}CO_2$

2M acid

outlet

Leaf area covered
with foil to
investigate light
requirements for
photosynthesis

Variegated leaf
to investigate
chlorophyll
requirement

$Na_2{}^{14}CO_3$

Figure 2 Bell-jar experiment for the investigation of $^{14}CO_2$
fixation in plants

whereas $10-20$ μCi will allow a more comprehensive series of investigations to be made. Incidentally, the maximum amount of carbon-14 which may be used per class experiment in schools is 640 μCi; far more than is required.

After the period of exposure to $^{14}CO_2$ and light, the plants are removed from the bell-jar. Depending on the amount of activity used, the method of removal will differ. With activities up to 1 μCi (see D.E.S. regulations, 1965), the unused $^{14}CO_2$ may be released directly into the atmosphere. By using higher levels in the experiment, the release of activities in excess of the permitted 1 μCi of gaseous waste

Figure 3 Print of an autoradiograph of a Coleus leaf exposed to $^{14}CO_2$

per day is likely to be exceeded. The $^{14}CO_2$ is pumped through a solution of caustic soda before the plants are removed. To examine the distribution of ^{14}C, samples of the various parts of the plant may be removed with a cork borer, dried and counted under a Geiger counter. Alternatively, visualisation of the distribution of activity may be made by using the technique of autoradiography (fig. 3), that is, placing a well-dried leaf sample directly against the emulsion of a sheet of X-ray film and leaving for a period of time for exposure to the beta radiation, which interacts with the emulsion to generate a developable image. With the higher activity levels, this exposure period may be a few hours, but for the 1 μCi levels several days may be required.

So far the carbon atom has been traced into specific areas of the plant. The next stage of the investigation is to determine the form in which the carbon exists. The plant material is extracted with 70 per cent ethyl alcohol and after concentration of the extract, the presence of soluble sugars may be examined by paper or thin-layer chromatography. Again autoradiography may be used to localise the radio-active spots and their identity is then determined by reference to stained inactive standards. The tracing of inorganic carbon into an organic form is now complete. The translocation of these labelled sugars throughout the whole plant system may now be investigated by selectively labelling only one leaf or side-branch. The distribution of the labelled sugars may then be easily traced using a portable Geiger counter. As a further extension of these experiments, the respiration of plants may be measured by confining a previously-labelled whole plant to a darkened bell-jar with a small container of sodium hydroxide. The release of $^{14}CO_2$ may be measured by observing the increasing activity of the hydroxide. To accomplish this measurement, the $Na_2^{14}CO_3$ formed must first be converted to $Ba^{14}CO_3$ and the solid is then dried and counted under a Geiger counter. The reason for the conversion is that if the $Na_2^{14}CO_3$ were dried and counted directly, over 60 per cent of the activity would be lost by exchange with atmospheric carbon dioxide. This is less of a problem with $Ba^{14}CO_3$.

Investigations involving animals.

The previous series of experiments allows the cyclical pathways of photosynthesis and respiration to be measured without difficult experimental procedures. Numerous investigations may be undertaken with plant material, mainly because of the relatively few handling difficulties. Generally, physiological experiments with animals are possible only with the invertebrates. In this respect, the locust (playing a prominent part in school biology teaching) may be used. The next experiment is a progression from the previous photosynthesis experiments to animal respiration.

By labelling a few leaves of plants with carbon-14, killing them and feeding them to locusts, the respiration of the animals may be investigated. This may be accomplished by confining two or three animals to a small chamber through which air is drawn and monitored for radioactivity. The $^{14}CO_2$ is collected in 2M NaOH and precipitated at regular intervals as $BaCO_3$, using 1M $BaCl_2$. Each sample of $Ba^{14}CO_3$ is weighed and counted as a dried sample, spread evenly over an aluminium counting tray (22 mm diameter), called a planchette.

During such an experiment over a two-hour period, several $Ba^{14}CO_3$ samples will be collected on the planchettes, but each sample will probably have a different weight of solid. Because of these weight differences counts from each prepared sample cannot be correlated one with another. This is because the beta-particle from the carbon-14 is weakly penetrating; so differing weights of solid will cause differing amounts of absorption. This problem is referred to as *self-absorption* and may be demonstrated, and corrected for, in the following experiment.

By preparing a labelled precipitate of $Ba^{14}CO_3$ and plating out a range (10–80 mg.) of differing weights on about six planchettes, the graph of fig. 4 may be constructed; thus relating count rate from the dried sample against its weight. This immediately demonstrates the problem of self-absorption, for without it the graph would be linear, as indicated by the dotted line. The self-absorption graph flattens out with weight until, no matter how much precipitate is placed on the planchette, the count rate from the sample stays the same. The reason for this is that at a

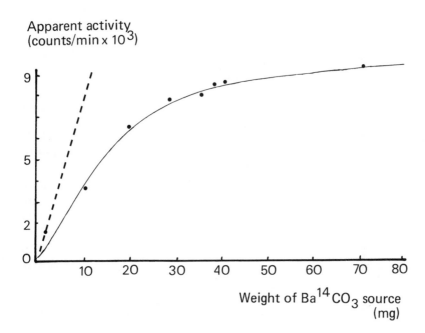

Figure 4 Graph of apparent activity of $Ba^{14}CO_3$
source against weight, illustrating self-absorption

particular weight or thickness it is only the upper layers of
the precipitate that are contributing to the count. The
beta-particles in the lower layers are all absorbed in the
precipitate.

Any count made on a sample will therefore only be an
apparent count. Knowing the weight of these individual
samples, the apparent specific activity may be determined
(i.e. apparent activity/weight). If this is plotted out, as in
fig. 5, then by extrapolating this curve back to zero weight,
where there is no self-absorption problem, the *true specific
activity* of the batch of six samples may be determined. This
is the true activity per unit weight of the sample. Using this
value and the weights of the six samples, the true activity of
each sample may be calculated. Because this is the true
activity of each sample, not the apparent activity, each of the
six samples may be correlated one with another. Unfor-

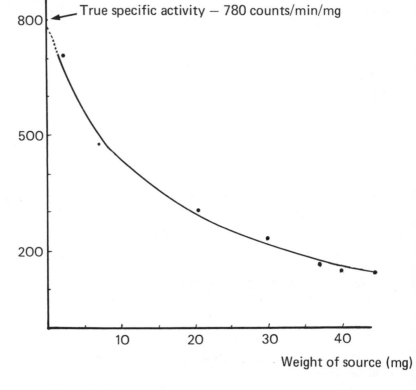

Figure 5 Graph of apparent specific activity against weight

tunately, this particular curve, and the true specific activity which was determined from it, only applies to the one set of readings. We want this information to be of general use, so that all sources may be easily corrected for self-absorption. To this end a final curve (fig. 6) is plotted, showing the percentage transmission against the weight of precipitate. This relates the percentage transmission of ^{14}C betas in a $Ba^{14}CO_3$ source to the weight of that source. To illustrate how this curve should be used to make the necessary self-absorption correction, assume that a prepared source from a respiration experiment weighs 15 mg. and gives an apparent count of 3,560 counts per minute. Reference to fig. 6 shows that at this weight there will be 52 per cent transmission of the beta-particles. The true count rate of the source will therefore be 6,846 counts per minute.

The isotope used in these experiments is probably the most commonly used biological tracer. The other isotope of biological importance is that of hydrogen (tritium) which has such a low-energy beta-particle that it will not penetrate the thin window of a Geiger counter. Detection of it is therefore extremely difficult and is only possible in schools using photographic emulsions and the more sophisticated

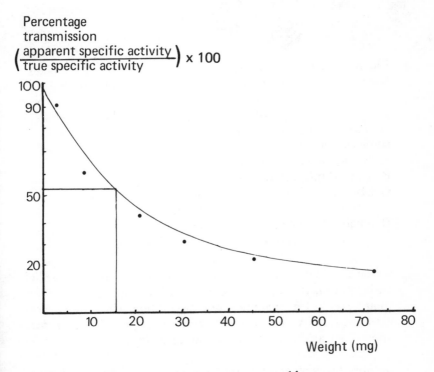

Figure 6 Percentage transmission in a $Ba^{14}CO_3$ source due to self-absorption

technique of *microscopic* autoradiography (Hornsey, 1970)
(fig. 7). At present tritium is not being used in schools but

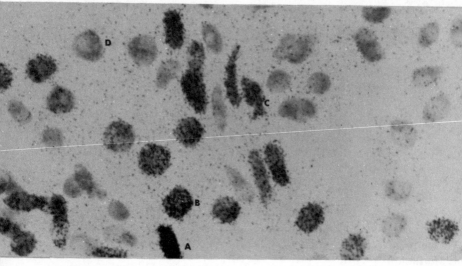

Figure 7 Micro-autoradiograph of root tip nuclei of the bean
Vicia that has been grown in a nutrient solution containing
tritiated thymidine. The black dots are developed silver grains
produced as a result of beta interaction with the photographic
emulsion. They can be seen to be closely associated with
some cell nuclei.
A heavily labelled nucleus
B labelled nucleus in interphase
C labelled nucleus entering division   —   individual
    chromosomes not easily identifiable
D unlabelled nucleus in interphase

may well be in the future in the form of *tritiated thymidine*.
This material is the pyrimidine base thymine plus sugar in
which the hydrogen atoms of the methyl groups are labelled
with tritium. By feeding this material to dividing cell systems
(plant roots, for example), the thymidine will be incorporated into DNA. The use of autoradiography will therefore
allow the tracking of chromosome divisions.

Biology teachers and laboratories
The use of radioisotopes provides an extremely useful
extension, and perhaps improvement, of some important

school biology experiments. At present, crowded time-tables appear to restrict most isotope investigations to pupils either undertaking project work in the third year of the sixth form, or in boarding schools where access to laboratories outside school hours is possible. With the advent of the Nuffield Advanced Level Science courses and the application of modern methods to biological investigations, radioisotope work will become a tool of the school biologist in the future, much as the microscope is at the present. For this reason it would be wise to encourage teachers to think in terms of using radioisotopes in practical classes. With such developments the question of special training and facilities is raised. The training required by teachers, in order to both work with and handle radioisotopes competently, is fairly minimal. The DES requires the teacher to have undertaken at least twenty-five hours of work involving unsealed sources of radioactivity. This work, usually in the form of a one-week course, will involve instruction in the safe handling of radioactivity, together with practice in experimental work. Such courses are designed to give the teacher confidence in taking the subject into his own school, and not only imparting that knowledge to the pupils, but also to the technicians who may well be involved in setting up experimental classes involving radioisotopes.

Most school laboratories may be used for radioactive work, providing a few precautions are taken. The most important aspect of this work is not the radiation dose from external sources but the *contamination* and possible *ingestion* of the unsealed source. For this reason an area of the laboratory should be set aside, which during a particular part of year, can be rapidly converted to radioisotope work. The benches should be covered with polythene sheeting so that any spillage could easily be removed and does not become fixed into inaccessible cracks in the woodwork. Work with radioactivity should also be conducted within the confines of a large enamel or plastic tray. Any spillage here would then be confined to this working area. A fume-cupboard would be an advantage, particularly in the case of photosynthesis and respiration experiments where radioactive gases are involved, but this is not essential. Good ventilation is the essential requirement.

Table 2 Suggested Experimental Work using Radioisotopes

Isotope/Form	Level of activity	Investigation	Additional Work
^{14}C as Na_2 $^{14}CO_3$	1 – 20 μCi	Photosynthesis, translocation, plant respiration	Effects of different wavelength light on carbon fixation. Environmental conditions for translocation.
^{14}C – substrates	2 – 5 μCi	Metabolism by bacteria, animals, plant cells by trapping evolved carbon dioxide.	Effects of inhibitors (CN etc), uncoupling agents (dinitrophenol) on cellular respiration.
^{32}P as orthophosphate ^{35}S as sulphate	10 μCi	Ion uptake (active and passive) by plant roots.	Effects of inhibitors, temperature and competitors.
^{22}Na as NaCl	2 – 3 μCi	Active transport across frog skin. Uptake by fresh-water animal.	Effects of inhibitors. Effects of animals living in varying salinites.
^{3}H as thymidine	5 μCi	Chromosome division in dividing cells of plant roots – micro-autoradiography	
^{131}I as NaI	1.5 μCi	Uptake of iodine in the thyroid of the tadpole – micro-autoradiography	

With the need for careful control of ingestion, no mouth operations should be allowed near the working area. Remote pipetting devices should therefore be used for both radioactive and non-radioactive work during a particular experiment. In the case of apparatus it would be advisable to have it reserved for radioactive experiments *only*. This would be particularly valuable in the case of pipettes, otherwise they may be used in experiments where mouth pipetting is allowed and accidental ingestion of the unsealed source might result.

In conclusion, I would like to emphasise that the use of radioisotopes in school work will require a minimal amount of equipment, but will bring extra sensitivity into practical experiments. When reasonable precautions are taken, the hazard from the radioactivity is negligible.

Table 2 lists the experiments I have described in this chapter, together with some other suggestions for practical investigations.

References

J. N. Andrews and D. J. Hornsey (1967) 'Experiments with radioisotopes in School Biology', *School Science Review 48*, 166.

J. N. Andrews and D. J. Hornsey (1972) *Basic Experiments with Radioactivity*, Pitman, London.

D. J. Hornsey (1970) 'Iodine uptake in the tadpole demonstrated by the use of radioactive iodine and autoradiography', *School Science Review 52*, 178.

D. J. Hornsey (1973) *Radioactivity and the Life Sciences*, Methuen, Nuffield Advanced · Level Science (1970), Teacher's Guide, *Maintenance of the Organism*, Penguin.

P. A. M. Paice (1968) 'Simple radioisotope experiments in School Biology', *School Science Review 49*, 170.

The Use of Ionizing Radiations in Schools, Establishments of Further Education and Teaching Training Colleges (Department of Education and Science) *Administrative Memorandum 1/65*, January 1965.

Science for all

DENNIS FOX

Very few of the pupils in schools today will become the scientists of tomorrow, and if science is to be justified as a necessary part of the curriculum of all pupils, it has to be justified in terms of *general* educational goals, and not in terms of training the nation's scientific manpower.

In 1965 the Nuffield Secondary Science team was charged with the task of providing material which would be used to achieve these general educational goals applicable to the vast majority of secondary school pupils between the ages of thirteen and sixteen. Because the Secondary Science Project was the largest of the science projects and the only one for this age group concerned with such a wide ability range, it is from the biological components of this project that I shall draw examples of the objectives and methods and problems and opportunities of providing 'science for all'.

Subject Matter

The guidelines for the project were set out in 1965 in *Science for the Young School Leaver*, the first working paper produced for the then newly-created Schools Council. 'The basis of these suggestions', said the author of the Working Paper, 'is that there are certain major themes in science, with which Newsom pupils might be expected to have some measure of acquaintance and, whilst it will be necessary to vary the manner and extent to which these are pursued... the themes themselves should largely be the same for all'. The cohesion of each individual theme and the overall pattern of the themes collectively was to give a significant underlying structure to the work.

Eight of these themes were listed in the Working Paper together with the 'principal emergent ideas and areas of knowledge of which there should be some appreciation'. This is how the subject matter of the Secondary Science Project came to be arranged in eight interrelated Themes. The titles of these Themes and the major interrelationships are illustrated in the diagram.

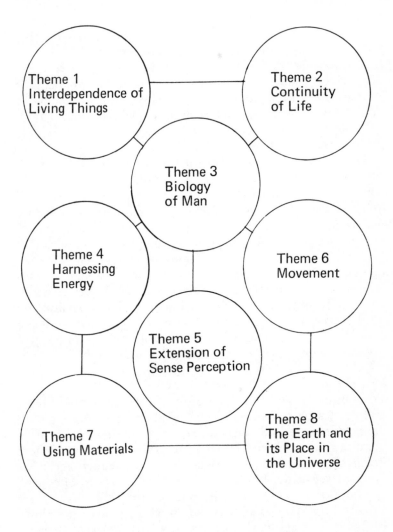

The eight Themes of the Nuffield Secondary Science Project

Getting the feel of being a scientist

A list of eight Themes, each with its associated 'area of knowledge', is clearly inadequate as a statement of the aims of a science course. Science is characterised by specific types of activities, as well as particular areas of knowledge, and the Newsom report of 1963 asked that pupils should 'get the feel of being a scientist'—not because they might one day become scientists, but presumably because there is something of value to everyone in the scientific way of doing things. 'Every day life brings the need to solve problems, to predict the consequences of actions and to evaluate the assertions of politicians, advertisers and scientists' (Misselbrook, 1971). The aim of the Secondary Science Project was to help people to do these things scientifically.

The skills to be developed for this are listed in Chapter 2 of the Teachers' Guide to the project. They are the conventional kinds of objectives associated with the skills of doing science that have appeared elsewhere with other materials. They are skills like accurate observation, experimental design, forming hypotheses, interpreting data, verbal fluency, literacy and numeracy. Some of these, such as experimental design, are objectives particularly associated with science; others, such as verbal fluency, are objectives common to all subjects but which nevertheless (perhaps especially) still need stating in the context of a science course. It is all too easy for these things to become nobody's business just because they happen to be everybody's business.

Significance

To consider the aims of the project in terms of the subject matter of science and the skills and attitudes associated with doing science is still inadequate. It is an inadequate view because it is subject centred; it starts with the subject 'science' and states what science is to be learned and what kinds of scientific skills are to be practised. These, of course, are important issues, but the pupil-centred view is at least equally important. This view starts with the lives the pupils lead and asks about the aspects of science relevant to these situations. It asks how science can contribute to the problems they face; how it can help them to find answers to the questions they ask.

This is what is meant by significance. The Working Paper foresaw that 'the most dramatic of the changes which the Newsom Report envisaged could well arise from concentrating upon significance in all aspects of the work, by using it indeed as a touchstone...' In fact it was used much more than as a touchstone; it was the main criterion upon which the material was judged at every stage of the trials.

It is a consequence of this pupil-centred view that they are asked to investigate fuels, weedkillers, textiles and building materials, and to examine data about the dangers of smoking, the morals of sexual behaviour and the ethics of advertising. These activities appear in the project not as fringe activities but central to the whole purpose of a *science for all* education. This has led to criticism that some of this material is not science and has no place in a science project. Such criticism misses the point because the question 'Is this science?' is less important than the question 'Is this relevant to the needs of the pupils, and if so what has science to contribute?' The subject thus becomes a means to an end, not an end in itself; education *through* science rather than education *in* science.

Selection of material
Because the Project was catering for such a wide range of pupils with varying needs and interests, in many different kinds of schools in different areas of the country, the material was never seen as a course to be followed but rather as a 'quarry from which teachers would select suitable material to build coherent courses' (Misselbrook 1971).

There is roughly twice as much material as most teachers might use in a three-year course, and in selecting a pathway through the material the teacher will rely more heavily on some Themes than others, but the course should be such that all pupils will have contact at some time with material from each of the Themes. It would be contradictory to the whole concept of the project to pick out the biological material and to teach it in biology lessons. It was the Newsom Committee which cast doubt on 'whether these divisions (into biology, physics and chemistry) have a natural relationship with the requirements of our pupils'. It might also appear inconsistent with the ideas of the project to single out certain components to illustrate new movements in the teaching of biology. Yet all teachers planning courses will start from the areas with which they are most familiar and use these as relatively firm bases from which to explore less well-known fields. What follows, then, is not an abdication of the ideal of integration, but a map of one of the bases.

Biological material is that included within the overall concepts of *life* and *living* and though this is mainly found in Themes 1, 2 and 3, there is a certain amount in some of the other Themes. These are Theme 6 in which the natural movement of living things is a significant component; Theme 4 which had close links with Theme 3 in dealing with the energy and power of human muscles; and Theme 5 where the limitations of the human senses forms the starting point for investigations in optics, acoustics and electronics.

Themes 1, 2 and 3 deal with different aspects of the major concept of life. Theme 1 is concerned with the *variety* of living things and their *interdependence*. Theme 2 is concerned with the *continuity* of life, in terms of variation and consistency from generation to generation and over the many generations of evolutionary time. Theme 3 focuses attention on one particular living organism and in a man-centred project that organism had to be man himself.

Investigations

It is important for these pupils to 'get the feel of being a scientist' said the Newsom Report and the whole spirit of the project is 'one of investigation in which first-hand observation and experiment provide essential opportunities for thinking and acting scientifically' (Misselbrook 1971). Biological material has certain peculiar features of its own which make important (and sometimes unique) contributions to a pupil's understanding of the principles of scientific investigation. These features are to do with the variability and complexity of the material and the multiplicity of variables operating in any natural situation. Biological investigations often involve pupils in careful observation and recording in the field, under natural conditions where it is difficult to control variables. This might lead to repetition of observations in different situations or experimentation in the laboratory. Pupils learn the importance of repeating experiments and making observations on several organisms. They come to appreciate the variability of living material and the importance of choosing the appropriate organism for a particular investigation. The concepts of sampling and control, and the problems of organising and drawing conclusions from multiple data, are not peculiar to biological investigations but they are more easily overlooked in other areas of science.

Field work

All these issues are central to much of the work in Theme 1, where the work starts and radiates from the observation of organisms in their natural habitats. Successive observations of a particular habitat will yield data about the growth of plants in that habitat. Some plants grow more rapidly than others, they grow at different times and in different ways. By setting up laboratory experiments, the factors affecting the growth of particular plants can be studied independently. By controlling other variables, the effects of temperature and different soil conditions can be isolated. The 'laboratory' of the biologist is not always a building and these experiments can be extended into the outdoor laboratory of the school lawn, where the effects of irrigation, simple fertilisers and commercially produced lawn stimulants on the growth of lawn grasses can be studied.

The concept of a community of interdependent plants and animals maintained in some kind of balance will emerge from the studies of natural habitats. Questions about how this community comes into being, and how the balance is maintained, can be investigated by controlled experiments where an area of the school grounds is cleared, or the balance is disturbed in some other way. This leads to a recognition that man maintains unbalanced communities, called crops, to provide for his needs, and maintains them by controlling pests and weeds. Investigations of the colonisation of broad bean by black fly, of cabbage by the large white butterfly, and of lawns by broad leaf plants can be made in the context of previous studies of colonisation in natural and experimental situations. Control can then be seen as something that can arise from man's understanding of the dynamics of communities. These examples from Theme 1 show a sequence of different kinds of investigations. *Observation* in the field is followed by controlled *experiment* in the laboratory which in turn leads to *application* of knowledge and techniques in the control of the environment. They also illustrate the emphasis placed on a first-hand study of *living* organisms. Biology is primarily the study of living things. The study of dead things and the study of pictures and models, and other people's descriptions of living things is valuable, because it often provides essential experience that cannot be otherwise gained, but teachers should beware of any tendency for this to replace the real thing.

Keeping a good supply of living material on hand for laboratory investigation calls for organisation and forward planning, and certain specialised facilities not demanded by other branches of science, in which the material can be taken from a drawer in March in exactly the same state in which it was put away the previous November.

Biology of Man

However not all living material presents the same problems. The most readily available living material in any school is the thirty or so specimens of *Homo sapiens* which colonise the laboratory all too spontaneously every time the bell goes. These are the specimens which form the basis for the work in Theme 3. Theme 3 is not the biology of man in general, it is the biology of particular men, or rather particular boys and girls. It is the biology of John, Roger and Susan and all the other members of 3C who have science on Thursday mornings. One of the aims of the work in this Theme is to encourage pupils to ask questions about themselves as living organisms and to equip them to search for answers to these questions scientifically. An example of this type of investigation comes from the work on physical activity. Physical activity was chosen as the starting point for physiological and anatomical investigations because of the close association with the central concept of energy, and because of the enormous variation in the energy turnover of the body that results from the activity of the muscles. This means that large and easily measurable variations in breathing, circulation and temperature regulation happen when the muscles are working. Nearly all children are naturally active organisms so they are not being asked to investigate anything artificial; they are just being encouraged to submit their natural exuberance to controlled quantitative investigation. To this end a number of situations are devised in which measurements of the force exerted, the work done and the power developed can be taken over varying spans of time. The most popular and useful of these is a specially-designed bicycle ergometer. This is a stationary bicycle which can be pedalled against varying loads for any length of time, depending only on the stamina of the subject. (Teachers have commented that it is the only method they know of producing physical exhaustion in a fit fifteen-year old in ninety seconds!) The pupils are encouraged to ask questions about what is happening inside themselves whilst their muscles are working. Amongst other less relevant (and sometimes unrepeatable) replies, they say that they get out of breath, they feel hot and their hearts beat faster. These effects can then be subjected to quantitative investigation. It is not enough for a pupil to notice that he gets out of breath; he is encouraged to ask *why* he gets out of breath, *how much* breath he uses and whether this is related to the amount of work he does. By collecting the breath in a plastic bag and measuring the volume he can discover this relationship. By means of a simple analysis of the air in the bag and comparing it with laboratory air he can note that oxygen is removed from the air by the lungs. If he can relate this to other occasions when oxygen is used, he might guess that it provides energy from fuel. If he is *told* (because there are limits to what a pupil can discover even in a Nuffield project) that one litre of oxygen will release about 19.2 kJ of energy from fuel, he can work out how much energy has been released whilst he pedalled the cycle. He already knows how much work he did on the cycle, and because the energy released will be very much more than the work produced, he now knows why he got hot and he can work out his efficiency, in the same way that he worked out the efficiency of a steam engine. He will probably find that he is more efficient than a steam engine, but not much more.

Levels of difficulty
This sequence of investigations is a very ambitious one and only a few of the more able pupils will be able to follow it

through all its stages. The important point is that the sequence will make sense however far a pupil gets. Some will get no farther than a measurement of power output and noticing some of the physiological effects of this. Others will get as far as measuring the air volume and relating the volume of air breathed to the amount of work done. Others will tackle the oxygen analysis but only sufficiently accurately to note that oxygen is removed by the lungs. Where material is used by pupils of widely ranging abilities it is important to give the abler ones something to get their teeth into and feel a sense of achievement, and yet at the same time to enable the slowest to feel an equal sense of achievement at their own level.

Most of the apparatus the pupils use in these investigations has been specially designed for easy operation and simple computation. How often have less able pupils been defeated, not by difficult ideas, but by a piece of equipment that was difficult to manipulate or by a calculation that was too complex? The dimensions of the equipment are such that they yield easy numbers. The cycle wheel is one metre in circumference, the bench surface is made up to one metre from the ground and, in a step-up exercise, the weight of the pupil is adjusted to a round number of Newtons by means of sand bags. Complicated percentage calculations are avoided in the gas analysis by using a technique which automatically starts with a 100 cm^3 sample. These devices have been criticised on the grounds that pupils are being denied valuable practice in arithmetic. There is a point in this, but where there is a danger that pupils will fail to get through to the science because they fall over the arithmetic, then surely such devices are justified.

Data at second hand

The importance of pupils getting the feel of being scientists, by making their own investigations, has already been stressed. The examples of work so far described illustrate investigations in which first-hand observation and experiment provide the core of the work. There is however a good deal of important science that is not accessible by first-hand investigation in the laboratory, and in any case, life is too short for pupils to discover everything for themselves. Any honest science teacher will recognise that well over 90 per cent of the science he knows has come from what he has read or heard from other people. It is dishonest to lead pupils to believe that unless they are involved in a frenzy of practical activity they are not doing *real* science. Second-hand science is not necessarily second class science. The ability to acquire, interpret and evaluate data from a variety of sources is just as important, and perhaps more important, than the ability to do experiments. In Theme 3, most of the work in human behaviour, and especially sexual behaviour, is obviously not suitable material for a laboratory exercise. Data on man's exploitation of his environment is not difficult to find, and much of it is controversial and contradictory, making it ideal for practice in evaluation and interpretation; but little of it can be obtained by the pupils at first hand.

Inheritance and evolution
It is with the major ideas of inheritance and evolution in Theme 2 that experience with data at second-hand is most evident. It is not too difficult to do genetics experiments with mice and flour beetles, and both these organisms find a place in the material, but the inherited characteristics of cows, hens, roses, and man himself, are of much greater importance. Data about milk yields, butter fat content and progeny testing of dairy cattle; data about the selective breeding of hens and the development of a new variety of rose all comes at second hand, and provides valuable practice for the effective presentation of data and its analysis, interpretation and evaluation.

One of the most demanding sections of the whole project from the teachers' point of view must be that on evolution in Theme 2. Experiments on evolution are virtually impossible, although there is one suggested investigation of the selective advantage of cyanogenic clover. How does one present evolutionary evidence that is at the same time extremely fragmentary yet all adds up to a meaningful story? Evidence about the evolution of man is presented in a series of fifteen line drawings of fossils or tools, together with sixteen large photographs and a number of plaster casts. Pupils examine this second-hand evidence with the aid of worksheets, much as they might, in other circumstances, examine a hedgerow or a sheep's heart. They are not doing experiments but they are nevertheless getting the feel of being scientists.

Behaviour and morals

There are a number of points at which work in the project is concerned with human behaviour. This most obviously happens in Theme 3, where initial investigations on the senses and perception form a starting point for limited excursions into such psychological topics as motivation, drives, frustration, attitudes, personality and the effects of drugs. In the section on human growth, the behavioural changes of adolescence are as important as the anatomical and physiological ones. There are sections in all the Themes concerned with questions of human behaviour in the mass. Man is viewed as a social animal and the behaviour of societies both affects and reflects individual behaviour. It is such behaviour that has so effectively exploited the environment to provide for human needs; it has also been the cause of much human misery. A study of human behaviour cannot avoid issues of morality, yet it is sometimes said that science cannot be concerned with morals. This is only true in the sense that science does not provide answers to moral questions. But it does illuminate moral issues, and it does provide both information and skills that can help individuals to build their own moral frameworks. It is an important function of education to help people to build, and live with, their own systems of morals which are consistent and workable within the normal constraints of society. The biology teacher is no less involved in this than his colleagues in other disciplines. This does not make him a preacher and he must resist any temptation to impose his own morality upon his pupils, especially in those areas, such as sexual behaviour, where there is no clear consensus within society.

Science for all?

Nuffield Secondary Science was developed for older pupils in secondary schools who would not take O-levels in science. Whilst this covers a large majority of the pupils in that age range it is not quite science for *all*. As long as there are different kinds of science for the able minority and the rest, the problems of selection will loom large. At what age is it to be decided that a child will do one kind of science rather than the other? If an O-level course is a five-year course, this implies a decision at eleven plus and this is one reason no doubt that many attempts have been made to adapt the O-level material to the needs of the less able. But this is an enormous tail being wagged by a very small dog. The normal O-level course, centred around a high level of conceptualisation, is a very restricted source from which to meet the needs of the masses. How much more sensible it would have been to have developed 'Science for All' first, and then transform elements of this on to a higher intellectual level for the O- and A-level scientists. Over the years we have had a lot of experience of offering watered-down academic courses to the majority; we have had little experience of what would seem to be the more educationally valid process of developing material for general use, and stiffening it up for the academically able. This would seem to be a much more promising approach and one to which we should move in the future.

References

D. Fox (1971) *Biology of Man*, (Nuffield Secondary Science, Theme 3), Longman.

J. E. Marson (1971) *Interdependence of Living Things*, (Nuffield Secondary Science, Theme 1), Longman.

Ministry of Education (1963) *Half our Future*, HMSO.

H. Misselbrook (1971) *Teachers' Guide*, (Nuffield Secondary Science), Longman.

Schools Council (1965) 'Science for the Young School Leaver', *Working Paper no. 1*, HMSO.

G. Wigglesworth (1971) *Continuity of Life*, (Nuffield Secondary Science, Theme 2), Longman.

Personal relationships

JANE JENKS

A new subject is appearing in school curricula—education in personal relationships. Sometimes it is religious education or general English in a new guise; frequently it is sex education in its most fundamental form, that is, reproductive anatomy and physiology. Increasingly it is becoming an inter-disciplinary and coordinated approach to education about human relationships, not only between the sexes but between parent and child, young and old. It encompasses not only physical facts about human relationships but also social and moral education in the widest sense.

Sex education, concerned as it is with the most interesting area of human relationships and the one most fraught with myths, guilt and anxiety, will still play a large part in education for personal relationships. Ideas and needs in sex education have changed and today objectives in teaching about sex under this wider umbrella need to be clearly thought out. A glance at almost any but the very latest of the books or other resources produced for use in schools gives the impression that, in the past, much sex education was aimed at the limitation of sexual activity and its discussion, and that although more recently it has been inspired by a genuine desire to help people, especially adolescents, in their sexual relationships, it has tended to be in terms of 'giving them the facts and letting them make up their own minds'. Both these approaches are unsatisfactory; the former may give rise to feelings of unnecessary shame and the latter to bewilderment. Young people are not willing to accept unquestioningly their parents' standards of behaviour, and society's shifting values provide no framework in which decisions about right behaviour can be made: consequently a new approach is needed.

A new approach—relevance

In secondary schools biology teachers are usually responsible for sex education, especially when it consists largely of learning about basic anatomy and physiology, often of the rabbit rather than man. Guidelines for a new approach, not only to sex education, but to science teaching in general were laid out by L. G. Smith in Schools Council Working Paper No. 1 *Science for the Young School Leaver* (1965). This was based on the recommendations of the Newsom report concerning the needs of pupils of average and below average ability between the ages of thirteen and sixteen. Of these pupils L. G. Smith says 'By the end of their school life, and many of them will do no science afterwards, much of what they have studied can have relatively little significance for them; it must often seem trivial or artificial to them and have little importance in the world as they otherwise know it or are likely to be concerned with it.' Having put forward significance as a basis for a new approach the paper goes on to point out 'that the problems of dealing with personal, social and moral implications when they arise in connection with science will have to be faced. No treatment of science with these boys and girls can, in fact, claim to be realistic or of adult stature if it excludes or sheers away from issues of this kind, and much that may be of vital interest to the pupils is likely to be in this category—very much at the interface of the traditional school subjects, all with their own vested interests.'

Moral education

In making science more relevant to the lives of pupils, science teachers will have to accept some responsibility for moral education. This does not mean that the teacher should moralise; he can provide information pertaining to moral questions so that, as far as they are able, pupils can make decisions which begin to take into account the various factors involved. This requires both teacher and pupil to distinguish between biological and moral questions, e.g. 'what are the stimuli which lead to petting?' and on the other hand 'how far should we go?' In addition teachers can help pupils to appreciate how much of their behaviour is not rational. Where moral questions are concerned teachers need help in determining criteria on which actions can be based, other than their own feelings and beliefs. Guidelines for moral education have been provided by the work of John Wilson and the Farmington Trust for Research into Moral Education. This does not aim to set out and justify a set of moral values, but tries to establish a set of principles or criteria on which we can evaluate the merits of a moral belief or believer. Wilson sees morality as a form of rationality which can be justified in terms of its effect on other people's interests. The morally educated person feels equal concern for others as for himself, knows the relevant facts, is perceptive to other people's needs and feelings, has the relevant skills and abilities to make moral decisions and is able to translate his decisions into action. The implications of Wilson's analysis for sex education are fully considered in *Sex Education in School and Society* (Dallas, 1972): this contains much discussion relevant to this field.

Sex education in the home

It has often been argued that responsibility for this type of education lies with parents. As a child first becomes aware of relationships between people, including sexual relationships, at an early age, the need for some sort of education about sex will arise in the home. While many parents would like to be able to help their children, they feel that they cannot do so adequately because they lack information or the appropriate vocabulary, or are simply too embarrassed. If parents are unwilling to discuss sexual matters then children get the idea that it is something that 'nice people don't talk about'. This can lead to anxiety and secretiveness, and leaves children vulnerable to other influences. Children are also confused when parents deliberately mislead them or are not honest in their feelings. They also become bewildered if they are given too much information before they are ready for it. Lack of information only serves to increase the children's interest in sex, especially as they are not oblivious to its prominence in the mass media. Children will always find ways to pursue this interesting topic, and in the absence of parents or another informed adult they will turn to less reliable sources of information, especially their peers. Whispered conversations in the playground are a notoriously poor method of gaining accurate information. Without acceptable channels of communication, and with incomplete and inaccurate information, many children feel frightened and confused about sex. Thus reassurance and lessening of anxiety is one important objective of sex education.

Whether parents will talk about sexual matters or not, a child learns from them basic attitudes to his body and its functions, and also about the roles and responsibilities of different members of the family. Any family, including one-parent families, families where the maternal and paternal roles are strongly differentiated or where one parent is dominant in child rearing, provides only a limited view of these roles and responsibilities; children need to be introduced to the variety that these can take in different family situations. There is a need for education by some outside agency in the responsibilities of marriage and family life, at least 'to break the vicious circle of the unwanted child poorly cared-for in mental and emotional health (although often physically healthy and materially well provided for) who tends to produce unwanted children in turn'. (Dallas, 1972).

Sex education in schools—the relationship between home and school

Until sex education in its widest sense can be transferred into the home, schools, both primary and secondary, must accept part of the responsibility for this aspect of their pupils' education. However, in this context the relationship between school and home is a delicate one and a careful approach is needed to obtain the confidence and cooperation of parents. They should not feel that the school is intruding into their relationships with their children or encroaching on their rights as parents. Usually parents are informed about the kind of information which is given or being planned, and they may be invited to the school for talks, discussions, or to see the types of methods and visual aids to be used in the classroom. When approached in this way, as described by one primary school headmaster (Chanter, 1966), the majority of parents are delighted and relieved that the school is taking the initiative.

This approach is more often used by primary schools, as specific education about sex is extended to younger age groups; in secondary schools sex education of one form or another is an integral part of most biology courses or may appear as part of a course about personal relationships, or general studies. Although parents have the right to withdraw their children from such classes, it is becoming more of an exception for parents to be informed about any specific teaching activity in this sphere. It is assumed that this is as much a part of the child's education as geography or mathematics, and will be taught in a responsible manner by a specially trained teacher in a carefully planned teaching situation. This can help to remove some of the special aura which has pervaded this area of learning in the past. No longer does it take place with a special teacher on special days in a special classroom.

Teachers must be aware, however, that it takes time for the conditioning of years to give way to a relaxed non-embarrassed attitude. Children may meet open discussion about sex, as opposed to the whispered secretive approach, for the first time in school. They may feel guilty and react in a very emotional way. This kind of behaviour has been used both to criticise sex education and as a reason for not providing it in schools, but 'one must ask why these children were left in such a vulnerable position not only to sex information in school, but also to the many sources of worry and anxiety out of school too; there is no excuse for ignoring the fact that unless *all* sex information is suppressed from *every* source, unprepared children are going to react emotionally to it'. (Dallas, 1972). However, presentation of the facts in a straightforward and unemotional manner shows a lack of sensitivity for the child's feelings. It is helpful to children if their parents are in agreement with, and show positive support for, the aims of the school in education about sex. The type of contact described by Chanter may give parents the confidence they need to be able to help their children more positively in future. It also helps to avoid the problems that arise when children receiving sex education at school attempt to involve their unprepared parents in discussions about sex. Another advantage of contact between home and school, particularly in a multiracial community, is that teachers need to become aware of the variety of local and cultural attitudes to sexual matters which must be respected. Unfortunately there will always be parents who would benefit from this sort of contact but will not be sufficiently interested to attend.

Sex education in the primary school
Among other factors, better health, better nutrition and earlier maturity result in more children reaching puberty while still at primary school. The idea that there is no need to provide specific information about sex before this time, apart from answering a child's questions naturally as they arise, is giving way to the practice of introducing children at an earlier stage to the basic concepts of reproduction, before the awakening of sexual emotions puts up a barrier of embarrassment. Young children at school are encouraged to explore the world about them. They are interested in themselves too and such interest will be stimulated by observing pet animals kept in the classroom, especially if they are allowed to reproduce. In this setting, children can begin to be made aware that although man is an animal, his ability to think and communicate makes human relationships quite different from those of animals. Even in the primary school sex education should be given in the context of family and society. Children need to be prepared for the changes that will occur in their bodies at puberty if they are to be spared misunderstanding and anxiety. Girls need to know not only the facts of menstruation but also practical advice on how to cope with it in the school situation. Boys need to be reassured that seminal emissions and the awakening of sexual feelings are a normal part of their development too. Both early and late maturing children can be helped if they know that there is a wide range of variation in the rate of human development, and that in this context 'normal' is a meaningless standard. In spite of these developments in primary schools, the secondary school teacher should not assume that the new class of eleven-year olds are a homogeneous group with respect to the type and quantity of sex education they have received.

New objectives and resources for sex education in secondary schools
The concept of sex education for older pupils is being

widened to include not only the provision of facts but social and moral aspects as well. What sort of practical help is available to biology teachers in secondary schools extending their concept of sex education in this way? In recent years a variety of resources have been developed to fulfil just this need, not only by biologists, but also in the humanities and in moral education.

Work on the fundamental biological concept of variation introduces the section on human reproduction in the latest version of the Nuffield O-level Biology course which is in preparation. A primary objective of sex education must be to replace common misconceptions with an understanding of the range of human variation. It is important in reassuring children approaching puberty of the normality of the changes in their bodies, and it is an important pre-requisite for both understanding human sexuality and in developing considerate attitudes to other people.

The Nuffield Secondary Science project (1971) is an integrated science course for mixed ability groups with a strong emphasis on the investigatory approach and experiencing science at first hand. It was developed around the recommendations set out in Schools Council Working Paper No. 1 (1965) and the principles behind moral education suggested by John Wilson are used where questions about the moral implications of science arise. Theme 3, 'Biology of Man', places human reproduction as part of the whole human life cycle including growth, development, maturity, ageing and death. A course on child development is also outlined in *Health Education—Patterns for Teaching* (Elliott and May, 1967). Venereal diseases are placed in their proper context in the field of study on health and hygiene, and an outline of the physical basis of behaviour is included. Sexual behaviour is not based entirely on conscious rational decisions and yet biological factors which affect behaviour are seldom taught in this context, e.g. the biological significance of the sex drive, behavioural differences between the sexes, and the powerful effects of the sex hormones. Details of the effects of the fluctuations in hormone levels on the behaviour of girls and women can also be found in Dalton (1969).

The idea of individual learning or small group work,

fundamental to the Secondary Science project, helps to provide an acceptable climate in which children, especially those shy and embarrassed, can communicate freely with the teacher. It removes the need for some special arrangement for asking questions privately, like question boxes or seeing the teacher out of lesson time, which require the appropriate written, verbal or social skills to operate usefully. Also, if teaching is to be effective, the teacher must use various techniques to evaluate the success of his pupils' learning and to receive feedback from them to enable him to be sensitive to their needs, especially where these differ due to variations either in the level of their knowledge or in their physical and emotional development. Working mainly in small groups, the teacher can assess these aspects on an individual basis. A course of independent learning through programmed work, providing basic information about human reproduction for eleven-year-olds working individually at their own pace, has been prepared by Reid and Booth (1971).

As well as communication between pupil and teacher, sex education aims to increase understanding and communication between people in general so that they can be sensitive to one another's needs and can act with other people's interests in mind; in other words, to promote what McPhail (in the Schools Council Moral Education project, 1972) calls 'the considerate way of life'.

Both the Schools Council Moral Education Curriculum project and the Schools Council Humanities Curriculum project set out to achieve these objectives through structured work with pupils. The Moral Education Curriculum project, directed by Peter McPhail, is based on Wilson's criteria for moral education. Materials developed by the Humanities Curriculum project consist of packs of words and pictures, including 'Relations between the Sexes' and 'The Family', drawn from a great variety of sources, and from which the teacher can select those which are both interesting and relevant to his class. Stenhouse, the director of the project, believes the basic approach should be one of neutrality. The teacher does not propagate his own views but helps pupils critically to evaluate a whole range of attitudes and moral opinions. Although this approach can be successful, it could be argued that the teacher is there as an educator and it is his responsibility to see that pupils are not in any doubt about which are the responsible attitudes to hold. Pupils generally value the opinions of teachers with whom they have established a relationship and are only frustrated by evasive replies. Teachers will also have introduced their own bias in the selection of material in the first place.

It is sometimes recommended that sex education (or information) is given separately to boys and girls, especially younger pupils. This approach makes sex education special and sets it apart from the rest of the learning carried out in school. It is an advantage when a class contains a mixture of girls already into puberty and later maturing boys, but as a general principle it leads to secretiveness and embarrassment between the sexes and stands in the way of the development of understanding and consideration for one another's feelings. It is still common for only girls to be given information about menstruation—presumably boys fulfil their curiosity about it in less desirable ways. How different is the approach of the two Schools Council projects in which the relationships between the sexes are fully considered. In the Moral Education project boys and girls can explore each other's needs and feelings in different situations, and learn to see the consequences of their actions for themselves and other people through a variety of imaginative methods. The Humanities project provides similar opportunities for exploring goals and expectations in marriage, and for gaining insight into the wide variety of relationships in marriage. Both projects also consider parental roles and responsibilities. This is supported by work on the needs of developing children, which pupils are encouraged to experience at first hand in study of the human life cycle, in Nuffield Secondary Science—Theme 3. By learning through experience, this work provides moral education for those children who find abstract concepts difficult to understand.

Another aspect of these resources is that by their very nature they help pupils to develop skills in the comprehension of written material, and through discussion, to develop verbal and other social skills. Skills in dealing with people, both face to face and over the telephone, and being able to fill in forms, are important pre-requisites for acting on particular moral decisions, e.g. getting contraceptive advice or visiting the VD clinic.

Vocabulary

One of the first objectives that sex education must achieve is to provide an acceptable vocabulary with which to discuss parts of the body and its functions—a vital factor in communication with doctors, counsellors, and other experts from whom help may be sought. Victorian reticence has left us with only common colloquialisms which, although often vividly descriptive, are considered 'dirty words' and have a consequent shock value in usage, particularly in the context of home and school. Usually the number of medical terms to be learnt can be kept to a minimum if simple descriptive terms are used instead, e.g. egg tube for oviduct or Fallopian tube. This approach is used in *How Life Begins* (Reid and Booth, 1971), which provides basic sex information for eleven-year-olds.

Questions about sex

Given the opportunity children will ask a wide range of questions about sex: a long list of frequently asked questions is included in the pamphlet produced by the Gloucester Education Committee (1966), and Alan Harris's two books (1969) provide a wide range of questions, and background information relating to them, for structuring discussions. Sometimes even young children will ask questions which imply quite a sophisticated level of understanding about such things as contraception and VD. The child may only be able to appreciate an answer in the simplest terms, but it is possible to teach biological concepts at a variety of different levels for pupils of different ages and abilities.

One question that is often asked is 'Why is there so much fuss about sexual intercourse?' Generally it is not possible to make the complex emotions of an adult sexual relationship simple and understandable to a child, or even to an adolescent who has not had a similar experience. Teachers should not feel they are failing their pupils if they have to say to them that they will understand when they are older. With adolescent pupils the unit on 'Making Love', a collection of love poems from the Schools Council General Studies project (1972), may be helpful. Alternatively, a good description of the physical process can be found in Desmond Morris's *The Naked Ape* (1967), but pictures or films, such as Martin Cole's *Growing Up*, which show people having intercourse, are less desirable as they establish norms while excluding the great variety which exists in reality. The same problem arises when teachers are pressured by their classes and feel it is their duty to describe their own experiences: this can be the beginning of myths about human sexual performance which do not take into account the range of human variation. In addition, personal privacy should be respected; privacy of teachers, parents, and the pupils themselves.

Evaluation of resources and visual aids

Learning through discovery, which is the basis of much modern science education, is not an appropriate method for some aspects of sex education. Consequently pupils will have to deal with quite a lot of second-hand data, and teachers will have to select quite a wide range of visual aids for use in their courses. There are certain criteria which should be borne in mind when evaluating these materials for use with particular classes. Diagrams in books and films generally use two-dimensional visual symbolism which, although commonplace to teachers, is difficult for children to understand. Often no indication is given of scale or position in the body, and inappropriate or misleading colours are used. Children interpret such diagrams in literal terms as this example shows: 'After seeing a film loop, a description of menstruation given by one girl which was agreed with by most members of her class, "Well, you see there's this ping-pong ball which rolls down the tubes and rips out the lining of the uterus".' (Dallas, 1972). Helping children to make three-dimensional plasticine models aids them in interpreting diagrams and is a good way of evaluating the level of comprehension of the diagram in the class. However, even models differ in significant ways from the structures they represent and one aspect of this work is to consider how any model is not like the real thing.

Films can be very useful as a means of providing basic information, as reinforcement of work previously carried out, or even as a means of evaluation if classes provide their own commentaries. They are useful in structuring education about

more abstract concepts concerned with personal relationships, e.g. education for marriage. Quite complex ideas can be
presented in a visual way which would otherwise be quite
difficult to impart to those with poor reading and poor skills
of comprehension. Other films can help people to learn social
skills which may help to relieve fears and anxieties, e.g. by
making them familiar with procedures in VD or family-
planning clinics, or as in *Preparing for Sarah* (Eothen) with
prenatal tests and examinations. Films are not an opportunity for 'teacher's rest' nor should they be used as a visual
lecture (given under a cloak of darkness) in place of sex
education in its widest sense.

Films should be used as an integral part of the learning
situation and to fulfil a clear objective. Some films about sex
include too much superficial information to be taken in at
one time and satisfy no other objectives. It is often not clear
for what type of audience a film was made: *Growing Up* is an
example of this type of film. Teachers should always preview
films to see if they are suitable for a particular class. The
most useful films are short single-concept films presenting
new materials or situations, which are not readily available to
the class by other means.

Some evaluation of resources is being carried out for
teachers. The ILEA Media Resources Centre has produced a
pack, 'Health Education—Guide Lines to Materials', and the
South London Working Party on Education about Sexually
Transmitted Diseases (the members of which are mostly
teachers) have recently evaluated the existing resources and
visual aids used in this area. They have also produced
worksheets and picture cards for stimulating discussion.
Local authority health education departments may also have
a resources centre which produces materials and visual aids
for use in schools, as well as information and help for
teachers.

Responsibility for sex education in schools:
selection of teachers
How far are biology teachers expected to take responsibility
for this new wider concept of sex education? Obviously it is
up to the teacher to choose his own objectives in this area,
but it is becoming increasingly difficult to limit teaching in

biology to the provision of factual information only. Teachers may well worry that they do not have sufficient expertise in these new areas to feel confident about teaching them, but now resources and materials are becoming available and teachers can select from them, using their own experience of their particular class's needs, abilities and interests as their guidelines. Short in-service training courses run by the education authorities can provide valuable guidance in using these new resources and approaches.

Ideally sex education is not just the preserve of the biologist, but an area for interdisciplinary study which could also involve teachers of English, religious education, home economics, physical education, social studies, history, and geography. Where this approach is used an essential feature is consultation between different departments. Pupils do not want to 'do', say, boy-girl relationships in biology, religious education and English unless each presents a different angle integrated with the others into a broad perspective. However, not all teachers, even biologists, feel able to deal with all aspects of human sexuality with their classes. A teacher's values and views about sex will fundamentally affect his thinking and no good can come of a teacher struggling on against his own embarrassment through a sense of duty.

In spite of this the responsibility will often fall only to the biologist; if he does not provide some sort of help there is no guarantee that others will. Sometimes outside specialists, e.g. nurses, doctors, health educators, have been invited into schools to deal with particular aspects of sex education. Some classes respond better in the freer atmosphere provided by the outsider; other teachers may find that their classes may not be willing to put questions to someone with whom they have not built up a relationship first. In this case the role of these specialists lies in providing information and resources for teachers, and in supporting teachers who are acting as informal counsellors to their pupils, when they meet problems outside their area of expertise.

Generally teachers select themselves for this kind of work through their own interest, and feelings of responsibility generated by contact with young people; indeed teachers have to be highly motivated to even attempt to become familiar with the large range of resources now available in this field—no mean task in itself. Some local authorities, e.g. the Gloucestershire Education Committee (1966, 1971), run courses to select and train teachers who have been recommended by heads of schools for this type of work. Group discussions aim to help participants to clarify their attitudes and they are encouraged to assess themselves as to their suitability for this work. It is likely, however, that 'A variety of teachers, young and old, mature and immature, demonstrating a variety of attitudes, methods and roles, both communicative, informative and authoritative, provides the ideal situation for sex education in its widest sense rather than searching for any one ideal teacher.' (Dallas, 1972).

Schofield, in his study of the sexual activities and attitudes of young people (1968), found a lively demand for sex education which was not being fulfilled. In his second survey (as yet unpublished), he interviewed the same people seven years later and found that they were much more aware of the inadequacy of any previous sex education they had received. At last the concept of sex education has been extended beyond simply providing facts to cover a wider range of objectives, including behavioural, emotional and moral ones. It has positive aims beyond the prevention of VD and illegitimate pregnancies, including educating people for stable marriage and responsible parenthood. Children brought up in a secure and loving family, trusted, regarded and appreciated by their parents, are more likely to grow up with the kind of self-esteem which helps them to respond sensitively in their relationships with other people (Dominian, 1972). Until sex education in the broadest sense can enable today's children to communicate with and respond to their own children satisfactorily when they become parents, teachers must decide what part they can usefully play in educating their pupils for the considerate way of life.

Summary of today's objectives in education
for personal relationships
1 To provide factual information on aspects of the whole human life cycle, from the beginning of a human life and including birth, child development, adolescence, maturity, ageing and death.

2 By providing information and reassurance, to remove feelings of anxiety, guilt and shame about sexual matters.

3 To open channels of communication between teacher and pupil, parent and child, and between the sexes.

4 To promote the considerate way of life.

5 To educate for parental responsibility and family life, so that children in future generations may grow up with the kind of self-esteem which will help them in their own personal relationships.

6 To give some insight into human behaviour, so that people may be aware of the non-rational factors involved in sexual behaviour.

7 To give some idea of the range of human variation.

8 To provide as far as possible a simple and acceptable vocabulary for parts of the body and its functions.

9 To help people to learn practical social skills, such as skills in written and verbal communication, practice in making decisions and the relevant skills for acting on their decisions.

10 To prepare children for changes at puberty, including changes in feelings and behaviour.

11 To prepare future parents so that sex education in its widest sense can begin at an early age and in the home.

References

A. G. Chanter (1966) *Sex Education in Primary Schools*, Macmillan.

D. M. Dallas (1972) *Sex Education in School and Society*, NFER.

K. Dalton (1969) *The Menstrual Cycle*, Penguin.

J. Dominian (1972) 'The End of Sex', in *The Tablet*, 25 November 1972.

D. S. Elliott and E. T. May (1967) *Health Education—Patterns for Teaching*, Macmillan.

Gloucestershire Education Committee (1966) *Education for Personal Relationships and Family Life*, Second Report.
    (1971) *Education in Personal Relationships and Family Life*, Third Report and Handbook.

A. Harris (1969) *Questions about Sex*, Hutchinson.

ILEA Media Resources Centre (1972) *Health Education: Guide Lines to Materials* from Highbury Station Road, Islington, London N1 1SB.

D. Morris (1967) *The Naked Ape*, Jonathan Cape.

Nuffield Secondary Science (1971) *Theme 3—Biology of Man*, Longman.

D. Reid and P. Booth (1971) *How Life Begins*, Heinemann.

M. Schofield (1968) *The Sexual Behaviour of Young People*, Penguin.

M. Schofield (forthcoming) *Sexual Development*, Allen Lane.

Schools Council General Studies Project (1972) *Family* and *Science and Responsibility*, Longman, Penguin, obtainable from publishers at 9—11 The Shambles, York.

Schools Council Humanities Curriculum Project (1970) *Relations between the Sexes* and *The Family*, Teachers' Kits, Heinemann.

Schools Council Moral Education Curriculum Project: Lifeline (1972) *Moral Education in the Secondary School*, Longman.

Schools Council Working Paper No. 1 (1965) *Science for the Young School Leaver*

South London Working Party on Education about Sexually Transmitted Diseases (1972) *Results of Inspection and Production of Resource Material*, Duplicated copy obtainable from Mrs D. Dallas, Faculty of Education, Kings College, London.

Films

M. Coles, *Growing Up* from Global Films, 143 Wardour Street, London W1.

Eothen, *Preparing for Sarah* from Eothen Films.

Further Reading

R. Rogers (ed.), (1974) *Readings in Sex Education*, Cambridge University Press.

Attitudes and motivation

JACK DUNHAM

'Sometimes we teach the beauty and importance of a subject as well as the substance of it. Sometimes, though, we teach people to dislike, and then to avoid, the very subject we are teaching them about' (Mager 1968).

One of the most difficult problems facing the teacher who is planning to use the approaches discussed in this book, is the change he should make in his attitudes towards his role. The function of the biology teacher, when the accumulation of knowledge is the primary, and perhaps the only aim, is quite different from his function when 'the learning of a method of thinking is a primary objective' (Falk, 1971). The most significant aspect of this change, Falk argues, is that 'instead of a purveyor of information he now becomes a guide in helping students uncover information for themselves'. Many teachers of biology would disagree that their role has been merely concerned with the communication of a store of facts. They may already have begun in their teaching to anticipate the changes discussed by my co-authors. If they have, they will already have started to consider the implications of these developments for the success or failure of their plans.

The most essential of these is that much more attention needs to be given to the factors which influence, and sometimes determine, the learner's response to teaching. 'We have to acknowledge that the learner is not merely a cognitive entity; on the contrary, he exhibits affective or emotional states in his motivations, attitudes, personality, anxieties, intentions and beliefs—all of which influence his learning in both direct and indirect ways' (Ausubel and Robinson, 1969). These cognitive and non-cognitive variables interact with each other. For instance, motivational factors can energise the process of learning by promoting attention and effort, on which the 'new' biology teaching is so dependent.

This interaction may be seen quite clearly in the study of the attitudes of pupils towards teaching and learning, and so the present chapter discusses the structure, formation and function of attitudes. It examines the problems of changing them and the methods which have been proposed to develop favourable attitudes to biology.

The structure and formation of attitudes

A pupil's attitudes may be thought of as his consistent ways of anticipating, evaluating and responding to people, events, and objects. These attitudes have important motivational qualities which can lead the pupil to seek out (or avoid) the people, objects and experiences with which they have become associated. Attitudes therefore have an important place in an interactionist perspective of teaching and learning, which proposes that a pupil selects from and processes the information in his environment according to his ideas, values and feelings as well as his concepts.

A three-part structure of attitudes has been proposed by social psychologists. The one put forward by Krech *et al.* (1962) consists of a cognitive component, such as beliefs about an object, a person or a situation; an affective or feeling component; and a response component. In this formulation a pupil can have a belief about some aspect of biology; for example, that the use of live animals in laboratory experiments is wrong. This can be accompanied by feelings of revulsion and a determination not to go near the laboratory again, until this part of the course is finished. Sometimes the response affects the whole of the biology course and not just a part of it. One student even asked to transfer to another course. In her laboratory work she was required to handle live animals and then kill them herself in order to continue the experiments. She was sickened by this expectation, which she could not accomplish. Her beliefs and feelings were so strong she could not cope with them. If this was what biologists did, she told her university counsellor, she did not want to be one.

The discussion of the structure of attitudes leads to the question of the formation of attitudes. Social psychologists have argued that attitudes are learned: there are several perspectives in the study of learning which are relevant to an understanding of the process of attitude formation. The experience and response of the student on a first-year university biology course has already suggested that consistent ways of perceiving, believing, feeling and behaving may be learned by being *associated* with important experiences, charged with emotion.

Learning theorists, with different perspectives, have argued that attitudes are formed as a result of behaviour being reinforced by reward and punishment. If the desired response is reinforced by appropriate rewards whenever the pupil makes it, then the behaviour of the pupil can be gradually shaped towards the goals the teacher holds for the pupil's learning. The timing and spacing of the schedule of reinforcements will have important effects in determining the success of the learning process. The most influential psychologist to propose this explanation has been B. F. Skinner, who has taught for many years at Harvard University, using many practical applications of his 'operant conditioning' perspective in his teaching. These methods are also likely to be successful, he has argued, in the laboratory: 'In designing a laboratory course, if we keep an eye on the students' successes and the way in which they are spaced, we are more likely to produce a student who not only knows how to conduct experiments but shows an uncontrollable desire to do so' (Skinner, 1968).

The third learning theory, which has importance for understanding the process of attitude formation, is concerned with the effects of *modelling*. Children model their patterns of behaviour on their parents and, particularly when they are teenagers, on their friends. Teachers become models for some pupils and therefore they have the opportunities and responsibilities which this involves. They can model, by their behaviour, interest, curiosity and enthusiasm for biology. They can also model, on the other hand, that they are over-emotionally involved in biology or that they are objective and critical about its claims, for example, the contribution which this science might make towards an understanding of human behaviour. The challenge of the modelling perspective for teachers has been simply, but effectively, expressed by Mager, 'we must behave the way we want our students to behave'.

The fourth approach to learning theory, to be discussed here, is concerned with the development of the pupil's *self-identity*. This approach may be illustrated by considering the contribution which the school can make to the formation of unfavourable attitudes towards biology. This may be caused by repeated failure, which leads to the self-concept, 'I'm no good at biology', followed by a withdrawal from that threatening situation. There may be a negative evaluation of biology as a result of a negative evaluation of the teacher, because of his personality, behaviour or even appearance. The pupil concludes, 'If that's a biologist, I don't want to be one'. Sometimes this withdrawing behaviour occurs as a consequence of the other students on the course. One student on an engineering course at university had this experience. He was isolated on his course because the other students thought him odd and peculiar. They formed this perception because he wanted more from life than obtaining a degree, a good job, getting married and owning a car. His hair was much longer than anyone else's and he disliked wearing a shirt. He was interested in architecture and liked to listen to poetry, written and read by other students in the university. His motivation for engineering became very weak, as he said, 'If the other students on my course are going to be engineers I don't want to be one' (Dunham, 1973). This withdrawing behaviour may also occur as a result of information presented in lessons and text books. Children may be told (or read) that 'Human reproduction should be studied as objectively and factually as the reproduction of the earthworm. We do not think of man as something apart from the rest of the living world'. For many of them, man *is* set apart from other animals because he writes music and poetry, paints, has a self-concept, a mind and even a soul. For children with this view of human beings, the biology teacher's 'scientific', objective, impersonal view of man will not be acceptable. They will refute his claims for his field of study and remain unmoved by his arguments for the relevance of its processes and products.

It should also be noted, and this is very important for the teacher, that attitude learning tends to be consistent.

Attitudes fit together and synchronise with each other, so that within a pupil or a teacher there is a system of attitudes in which the parts are in equilibrium. Fresh learning may bring disturbance and dissonance, and the attitude system may have to change. On the other hand, the system may be so strong and closed that a fresh stimulus, which might be disturbing, will not be accepted by the teacher or pupil, even for preliminary consideration. Rokeach (1960) in his book, *The Open and Closed Mind*, has shown how novel ideas and experiences, which might have been expected to be change-agents, are blocked-out by these defensive, closed, interlocking systems.

The function and modification of attitudes
The study of the formation of attitudes can now lead to a discussion of the function of attitudes. It has already been proposed that one major function is to protect the individual from information and experiences which are potentially disturbing and possibly threatening. This can be described as a negative, blocking, and inhibiting function. It was called by Katz (1960) 'ego-defensive'. This is the aspect of attitude function to which many teachers refer when discussing their teaching problems. Many of their difficulties with pupils, parents and colleagues appear to be related to attitudes functioning in this protective way.

It is important therefore to argue that there is another function of attitudes, which is positive, energising, and facilitating. It might be called an 'ego-growth' function. This is related to the opportunities, which an individual's pattern of attitudes allows, for expression of his own values. This pattern provides a framework for the individual to seek (through knowledge) an understanding of his environment, so if learning is seen as facilitating personal growth, it will be approached eagerly.

Often the biology teacher is not working with eager, enthusiastic pupils, but with children who would rather be elsewhere, doing other things. Their attitudes may range from neutral to unfavourable. In these circumstances the teacher may wish to move the pupils' attitudes along the continuum towards the neutral, or even the favourable, pole. The study of the functions of attitudes should therefore lead to a discussion of attitude change.

Attitude change is much more complex than is sometimes assumed. It is important to recognise this so that a realistic appraisal can be given of the contribution which a teacher might make to such changes. The discussion of three questions will show the need for caution. These are: What is known about the process of attitude change? Have any conclusions been reliably established as a result of research studies? What is still only conjecture?

Teachers frequently act as if they believe that attitudes can be changed merely by providing new information, or even by exhortation. Research studies have indicated, however, that attitude change depends upon several factors. Three of the most important of these are: the perception of the person presenting the information; the form in which the information is given; and the characteristics of the people who are receiving the information. Thus one of the important variables appears to be the credibility of the communicator, in this case, the biology teacher. If he is perceived as an expert who is 'fair and trustworthy', his presentation of the field of biology is more likely to be acceptable to his pupils than if he is not. The pupils' perception of the teacher is very important in this communication process. It may be influenced by factors outside the teacher's control, for example, by the pupils' previous experience of biology teachers, or by the predominant attitude towards teachers in the catchment area of the school.

The importance of group influences on a person's ability to modify his own attitudes seems to be so well supported by research evidence that it might be proposed as an established principle. One research worker, after a thorough literature survey in connection with his own investigation of the effects of television on attitudes, concluded: 'It is true to say that expectations of social approval and disapproval in relation to the group play an important part in deciding whether or not there will be an attitude change' (Halloran 1967). This suggests that group norms, which pupils' groups establish in order to regulate the behaviour of their members, will determine to a considerable extent the response of individual pupils to a teacher's presentation of information, plans for experimental work, choice of topics for discussion, questions to be answered and new methods of learning.

The importance of the third question—What is conjecture? may be shown by referring to the question of the use of fear-arousing communications in the modification of pupils' attitudes. This question may be raised when the teacher is considering how behaviour might be changed in the laboratory, for example, to bring about a better observance of safety regulations. It might also be raised when important issues related to pupils' health are being discussed, for example, the question of the relationship between cigarette smoking and the later development of lung or throat cancer. Many teachers behave as if they hold the view that fear-appeals are effective in changing both attitudes and behaviour, and that the stronger they are the greater will be their effect. On the other hand there are teachers who propose that emotional messages of this kind are ineffective, because of the resistance they generate in their pupils. They argue in support of rational appeals rather than fear-arousal messages. They believe strongly that attitudes can be changed by the presentation of factual information. All these questions are still being investigated. The answers so far provided by research are tentative. If it is true, as Halloran (1967) has suggested, that 'It is unwise not to recognise that we have many conjectures but few facts' with reference to attitude modification in general, then his warning will be particularly relevant to the work of the biology teacher. Great care should be taken with fear-arousing appeals because their influence may be the direct opposite of a· 'common-sense' expectation. It is not true, for instance, that relating cigarette smoking to cancer will necessarily reduce the numbers of young pupils who are heavy smokers. The issue is not quite so simple or straightforward as Falk appears to believe: 'No opportunity should be missed to describe the results of research which show so clearly the definite cause and effect relationship between cigarette smoking and the crippling and killing diseases of lung cancer, emphysema and circulatory disorders. One of the most impressive ways to demonstrate this to a class is to contact a local county hospital and obtain the lungs from a recent living cancer victim which clearly show the diffusion of the cancerous tissue in the black "tarry" areas of the lungs. If students didn't believe it before, they will after they have seen it!' It seems incredible, but research studies have demonstrated that a strong fear-arousal communication of this kind may not have as much influence as a much less disturbing message, because of the 'ego-defensive' responses which such frightening information may evoke. (Triandis 1971). These responses may lead pupils to smoke more, and not less, as they attempt to control their feelings of anxiety, which, if unchecked, may become associated with many parts of the whole biology course.

This discussion of ways in which to produce attitude modification has given the impression of the teacher coping with a series of crises caused by unfavourable attitudes, which may develop at any time as a result of what she (or others) may do. Crisis-management-teaching of this kind may be an inevitable part of the teacher's role. But many teachers are not satisfied by this approach. They prefer a management-by-objectives approach. They believe that they can promote the development of favourable attitudes and they are alert to prevent the formation of unfavourable ones, because these may be difficult to change, as we have seen.

The promotion of favourable attitudes

Three approaches for promoting the development of favourable attitudes to biology in the classroom and laboratory will be examined. The first is the satisfaction of the pupils' needs. The second is the encouragement of approach behaviour. The development of scientific enquiry methods of learning is the third.

Teaching and learning situations provide many opportunities for the satisfaction of the pupils' (and the teachers') needs. Maslow (1954) has proposed a hierarchy of needs. At the base are those which are physiological. Next are the safety needs. If these are fairly well satisfied it becomes possible for the next need stage to be reached, 'Now the individual is aware of, and can pay attention to, his needs for love and affection and belongingness' (Cotgrove *et al.*, 1971). If these social needs are gratified, he can now begin to search for opportunities to satisfy his needs for individuality and identity. Finally, at the top of the hierarchy, when all the lower level needs have been satisfied, an individual is able to pay attention to self-actualisation or personal growth. The realisation by the pupil that learning biology can bring, as Ausubel and Robinson (1969) have argued, 'The thrill of discovery, a sense of greater knowledge, the warmth of a smile of recognition from a teacher who appreciates the pupil's attempted answer to a problem, means that biology has become not just a school subject but an important source of personal satisfaction'. These experiences may have important consequences for the development of self-identify. The statement, 'I am good at biology', may become 'I am pretty successful'. This may lead to the feeling that he could be a successful learner in the future. He will then approach these learning tasks with confidence and expectations of success.

The second approach to the development of favourable attitudes is based on learning theory. Mager (1968) describes it as 'Teaching according to the principles of approach behaviour development and minimising the arousal of avoidance behaviours'. In practical terms this means: 'When the student is in the presence of biology he should not also be in the presence of aversive, i.e. unpleasant conditions but he should be in the presence of positive or approach conditions. If biology becomes associated with unpleasant conditions, these circumstances surrounding biology may trigger off unpleasant associations and avoidance behaviours'.

There is a variety of behaviour which indicates avoidance: for example, no voluntary contact with the subject or the teacher of the subject; not talking about it; inventing an excuse to maintain distance from the subject or teacher concerned. These examples suggest that an important consequence will be that the pupil, by his avoidance behaviour, is unlikely to have the experiences which will modify these behaviours and develop approach tendencies. Mager asked sixty-five of his ex-students about their most-favoured and least-favoured subjects, in an attempt to discover the behaviour patterns of their teachers which had promoted approach, as well as avoidance, responses. Positive motivators were teachers who: 'Taught us how to approach a problem so that we could solve it for ourselves and so gave us the tools for learning. They also asked for and respected the opinions of pupils even though they did not always agree with them.' In addition, these teachers never allowed fear and anxiety;

humiliation and embarrassment, or boredom and monotony, to develop to any appreciable extent in their lessons. They never caused their pupils to think less highly of themselves; so their pupils never lost their self-respect and dignity. It is hardly surprising that these teachers, and what they stood for, were approached and not avoided.

The third perspective in the development of positive attitudes to biology is based on 'the use of scientific enquiry as the methodology of learning' (Falk 1971). It is argued that the traditional methods of teaching have emphasised deductive rather than inductive thinking. The new methods are based on the argument that inductive thinking, which is the method used in scientific research, will stimulate more positive approaches towards biology, because of its empirical and investigative characteristics. It is also claimed that these methods will encourage the development of the values which are necessary for scientific enquiry: 'Desire for knowledge and understanding, belief in scepticism and open-mindedness, demand for the verification of hypotheses, respect for logic, awareness of the implications and the consequences of scientific research' (Falk 1971). In essence, it is argued that these methods of learning biology will stimulate curiosity, exploration and involvement. By encouraging his skill and interest in learning, they will help the pupil to become a life-long learner, not only of biology, but also in his everyday activities.

If these changes in the ways in which teachers define their roles, *did* have these effects, then this inquiry or investigative method could be widely recommended. It is important, however, to note Falk's warning, that these claims are not yet supported by evidence: 'We do not know if the optimum learning for students occurs by discovery methods i.e. where they ask their own questions and creatively devise their own methods of solution or whether they learn better when teacher-directed methods are used' (Falk 1971). Caution against the uncritical acceptance and propagation of investigative methods and the rejection of directed methods has also been counselled by Ausubel and Robinson (1969). They are critical of enthusiasts for discovery methods, such as Bruner, who seems 'to perceive learning by discovery as a unique and unexcelled generator of self-confidence, of intellectual excitement and of motivation for sustained problem-solving and creative thinking'. They argue that the skilful exposition of ideas by a competent teacher 'can also generate considerable intellectual excitement and motivation for genuine inquiry'. After they had completed their appraisal of the relevant studies, Ausubel and Robinson reported that 'actual examination of the research literature allegedly supporting learning by discovery reveals that valid evidence of this nature is virtually nonexistent'.

There is some evidence from research studies, which have investigated the learning of university students, that with some learners discovery methods are inappropriate in that they result in confusion, apathy or antagonism and failure. These studies indicate that 'students, who have a strong need for direction and organisation, perform best in structured and formal learning situations' (Dunham 1973). It seems important and urgent therefore to investigate the relevance of this interactionist approach in secondary schools; in other words to carry out research in the classroom which attempts to show how the effectiveness of methods of teaching varies with the cognitive and affective characteristics of pupils.

Summary

1 Affective or emotional states may influence, or even determine, the learning which occurs in the classroom.

2 Attitudes to people, events and objects will play an important part in the interactions between teacher and learner.

3 Attitudes may be learned in several ways but the pattern of attitudes developed by an individual will tend to be consistent.

4 Attitudes may function in a protective way, leading to the rejection of new ideas. On the other hand, they may facilitate the acceptance of new ideas, particularly if learning increases personal growth.

5 The changing of attitudes is a complex process. Two of the more important factors which influence these changes are: the pupil's perception of the teacher; and the characteristics of the pupil and his group.

6 Teachers should be cautious in the methods they use in attempting to change their pupils' attitudes, for example, in

the use of fear appeals.

7 Favourable attitudes to biology have been promoted by (i) considering the pupils' needs, (ii) avoiding aversive conditions of learning, and (iii) using investigative methods of learning.

8 More research is needed into the learning processes which occur in the classroom, in order to examine the interaction between teaching methods and the pupils' concepts and attitudes. It should not be assumed that investigative methods are appropriate for all pupils, and so it is important that teachers develop realistic attitudes concerning the advantages and disadvantages of discovery approaches to learning.

References

D. P. Ausubel and F. G. Robinson (1969) *School Learning*, Holt, Rinehart and Winston.

S. F. Cotgrove J. Dunham and C. Vamplew (1971) *The Nylon Spinners*, Allen & Unwin.

J. Dunham (1973) 'Authoritarian Personality Traits Among Students,' *Educational Research*, in Press.

D. F. Falk (1971) *Biology Teaching Methods*, John Wiley.

J. D. Halloran (1967) *Attitude Formation and Change*, Leicester University Press.

D. Katz (1960) 'The Functional Approach to the Study of Attitudes', in *Public Science Quarterly 24*.

D. Krech R. Crutchfield and E. Ballachey (1962) *The Individual in Society*, McGraw-Hill.

R. F. Mager (1968) *Developing Attitude Toward Learning*, Fearon.

A. H. Maslow (1954) *Motivation and Personality*, Harper.

M. Rokeach (1960) *The Open and Closed Mind*, Basic Books.

B. F. Skinner (1968) 'Teaching Science in High School—What is Wrong?', *Science, 159*.

H. C. Triandis (1971) *Attitude and Attitude Change*, John Wiley.

Further Reading

J. S. Bruner (1961) 'The Act of Discovery', *Harvard Educational Review*, 31.

B. Z. Friedlander (1965) 'A Psychologist's Second Thoughts on Concepts, Curiosity and Discovery in Teaching and Learning', *Harvard Educational Review* 35 (1) 18–38.

B. Y. Kersh and M. C. Wittrock (1962) 'Learning by Discovery: An Interpretation of Recent Research', *Journal of Teacher Education*, 13.

L. S. Shulman and E. R. Keislar (eds) (1966) *Learning by Discovery: A Critical Appraisal*, Rand McNally.

E. Stones (1970) *Readings in Educational Psychology*, Methuen.

Approaches to assessment

ROBERT LISTER

At certain stages in a course assessments of some kind are usually required. Teachers need to know how successfully their students are learning and what points of weakness need to be reinforced. Assessment of this kind is really part of an on-going process of education. In it, the teacher is looking for mastery of the work up to a standard which he finds acceptable. In contrast to this, assessment for grading students aims at forms of testing which will spread the results widely so that the differences between good and poor students can be readily discriminated. Whether formal tests or assessment by personal impressions are used, only actual behaviours of students can be observed. Hopefully, we assume that these denote certain intellectual qualities which can be measured. The prudent teacher will attempt to define these qualities and then to express them in behavioural terms before designing a test. Some of these behaviours are mentioned later in the chapter in relation to examples of questions.

A range of assessment techniques is available. They may be represented on the following spectrum:

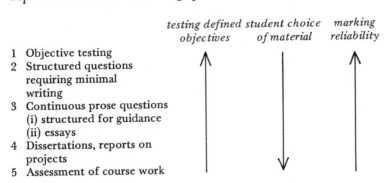

	testing defined objectives	student choice of material	marking reliability
1 Objective testing			
2 Structured questions requiring minimal writing			
3 Continuous prose questions (i) structured for guidance (ii) essays			
4 Dissertations, reports on projects			
5 Assessment of course work			

The two most desirable reforms in examining are in the

direction of greater precision and increased liberalisation. A combination of assessment techniques might produce improvements in both of these. Provided that they are of sufficient refinement and aim at testing particular attributes, they could be used to produce profiles of students which might be of more value than a single examination grade for predicting future performance (Lister 1969).

Continuous prose questions

This type of question is most familiar to biologists. Used in papers offering a choice of questions, they provide opportunities for showing that certain parts of a course have been studied in depth. In them, good candidates have scope to show their merits. On the other hand, the various choices of questions results in one rank order of candidates who have, in effect, taken different examinations. There is an inevitable variation in the level of difficulty of the questions and success is partly a matter of prudent choice by candidates. It is axiomatic that candidates use recalled information wherever possible. Questions requiring routine expression of revised material tend to be selected rather than those which require mental effort. Because of this and of the complex nature of the answers it is difficult to relate them to a discrete set of criteria. This makes marking unreliable.

Unless a question is well-structured, the application of an analytical marking scheme will fail to fit the variety of valid answers which it may produce. There have been several investigations which show that continuous prose answers graded after rapidly reading them to gain a general impression of their standard provides a more reliable assessment than using a mark scheme. The following question requires a true essay as an answer.

> Write a short essay on the interdependence of plants and animals based on observations made in any habitat you have studied.
>
> (Cambridge Local Examinations A-level 1954)

The most reliable method of assessing this would be for two or more independent markers to grade on impression and for their results to be combined. Such a combination can take advantage of the tendency of different markers to look for different attributes and thus a balanced view of each answer is likely (Wiseman 1961, Britton 1963, Head 1966). If certain responses are required, the question must be structured to provide guidance for the student. The answer will still require some organisation of prose writing, as in the following example:

> Describe the structure of the cells of the vascular cambium of a woody dicotyledon.
> Give an account of the development from the vascular cambium of (i) a wood fibre, (ii) a vessel member, (iii) a ray.
> Briefly indicate the differences in structure between the cells of the vascular cambium and (i) apical meristem cells, (ii) cork cambium cells.
>
> (J.M.B. S-level Biology 1967)

A well-defined marking scheme can be used for this. As it is mostly concerned with recall of particular information a strong element of objectivity can be used. It is not easy to allocate a proportion of marks for style in a mark scheme. There are widely different notions of what this implies and attempts to dissociate the medium from the message produce further difficulties. Provided that the logic and coherence of the writing communicate the information to the reader in a sufficiently meaningful manner, it can be assumed that the style has reached an acceptable level. This type of testing is imprecise for particular attributes other than the ability to organise a body of information into meaningful prose.

An examination consisting of a few large questions tends to discriminate between good and poor candidates less than does a test consisting of numerous small items (Ebel 1965). The sampling error of the course is compounded by marking error. Discrimination is further reduced by a general reluctance to award very high or very low marks. The results of the majority of students thus tend to accumulate in the middle of the range making discrimination between them difficult. In spite of these disadvantages, it may well be necessary to test the ability to communicate complex ideas even at some sacrifice of reliability in marking.

Objective testing

This includes several techniques, all of which require an answer which is pre-determined when the objective item is written. This gives reliability in marking, but constructing items is itself a highly subjective task and requires great care. Before using them in a main test they should be tested on a corresponding population of students for level of difficulty. Items of medium difficulty in which 30 per cent to 70 per cent of the class obtain the correct answer, and in which good students show a high success rate, and poor students a low success, are the most useful for grading purposes. Items can be improved and used again after a test. Information for making improvements can be obtained by noting the frequency with which different alternatives have been used. It is quite possible to develop a steadily improving bank of items which can be used for rapid testing during a course. Tests can include any of the following types of items:

1 *True-false items*   A large number of simple items can be produced to test factual knowledge. This can be useful for quick testing at the end of a lesson. With care, true-false items can also be constructed to test critical thinking in biology. Modifications of this type of item can ask for -true, probably true, insufficient evidence available, probably false, and false responses.

2 *Completion exercises*   The student has to provide a missing word to complete a sentence. This eliminates the chances of guessing from provided alternatives. Construction of such items is easy and they are therefore suitable for classroom tests. The sentences must, however, be designed to give suitable guidance to the desired answer or the teacher will be obliged to allow several alternative answers.

3 *Multiple completion items*   These use a letter code for selecting one or more of the alternative answers offered. This is fairly complicated and it is usually better to use the multiple choice form.

4 *Multiple choice items*   These are the most straightforward and popular forms of objective testing. A correct alternative has to be chosen from four or five alternative answers given. They are often used for testing recall of factual knowledge, but it is also possible

to use them for testing a range of skills used in biological science. Each item should aim to test one step in thinking.

For example, the following item tests knowledge of the correlations of factors affecting the blood system of the dogfish:

> Which of the following pairs of factors affecting blood circulation shows an inverse relationship?
> A blood pressure and metabolic rate;
> B heart rate and oxygen supply;
> C heart rate and rate of water flow over the gills;
> D the concentrations of oxygen in the blood and in the water leaving the gill system.
> Key D

On the other hand the following item, which was intended to test interpretation of numerical data, was ambiguous and consequently gave poor discrimination in a trial test.

> The following results were obtained when a seashore rock pool containing plants and animals was sampled for oxygen throughout the day:

Time of day	07.00	09.00	11.00	13.00	15.00	17.00	19.00
oxygen cm^3 l^{-1}	4.3	4.8	6.4	7.2	8.9	8.9	7.3

> Which of the following processes is the most likely cause of these changes?
> A photosynthesis;
> B respiration;
> C photosynthesis and respiration;
> D neither photosynthesis nor respiration (Nuffield A-level trial 1967).
> Key C

As the data provided insufficient information on which to make a decision, the item could not be improved and was rejected.

To produce a multiple choice item, it is usually possible to think first of a direct question. This becomes incorporated in the introductory statement (stem) which poses the situation with precision and economy. In producing the alternatives, mere incorrectness is insufficient. Each alternative must show a plausible relationship to the stem. The following possibilities can be used:

1 the opposite of the correct answer;
2 a common misconception;
3 a true statement which does not satisfy the requirements of the question;
4 a statement which is too broad or too narrow for the requirements of the question;
5 an incorrect response which appears plausible unless thought is exercised.

The difficulty of an item is influenced by the fineness of the differences between the alternatives. If this is too fine, we approach ambiguity. For example, the following item introduced too fine a logic for many candidates who were driven to guessing:

A Latin Square is the name given to an experimental layout which is:

A a statistical device that enables experimental errors to be eliminated;
B an experimental procedure that compensates for variations in factors which are difficult to control;
C an experimental layout that allows corrections to be made for random experimental errors;
D a method of planning an experiment so as to eliminate errors due to uncontrollable variable factors (Nuffield A-level trial 1967).

Key B

As a result of a trial test, the item was rewritten as follows:

A Latin Square is the name given to an experimental layout which

A ensures that experimental errors are prevented;
B allows the experimenter to take account of the influence of variations in the environment on the results;
C eliminates the effects of uncontrollable variations in the environment upon the organisms;
D reduces edge effects.

Key B

In this form, discrimination was considerably improved (Nuffield A-level trial 1967).

Matching exercises provide opportunities for relating examples to principles, structures to functions and data to interpretations. Each set of alternatives acts for each item in turn and thus has to function as a correct response and also as an incorrect distractor. It is not easy to produce such an exercise to give comparable difficulty and discrimination in each of its parts. The following exercise was intended to test the relation of each situation to a principle.

For each of the following numbered statements on photosynthesis, give one of the explanations A to E. You may use each letter once, more than once, or not at all:

Keys

1 at low light intensities, an increase in temperature
   has little effect on the rate of photosynthesis;   D
2 at a high light intensity, an increase in temperature,
   up to a limit of $35^{\circ}C$ causes an increase in the
   rate of photosynthesis;   E
3 above a certain intensity, a further increase in light
   has no effect on the rate of photosynthesis;   C
4 when the rate of photosynthesis is not limited by
   light, an increase in the concentration of carbon
   dioxide will increase the rate of photosynthesis;   C
5 when carbon dioxide concentration is increased
   beyond 1 per cent the rate of photosynthesis does
   not increase   A
because
A carbon dioxide acts as an inhibiting factor;
B light intensity depends upon temperature;
C carbon dioxide acts as a limiting factor;
D light reactions in photosynthesis are unaffected by
   temperature;
E the dark reaction is speeded up (Nuffield A-level trial
   1967).

B is a nonsense alternative which was little used. Items 1 and 5 were poor discriminators, item 4 was ambiguous (E was also a possible answer) item 3 was very good.

In contrast to this trial exercise, the following one gave uniformly excellent discrimination. It also shows the use of diagrams and tabulation in an objective test:

The figures A to D show different pyramids of numbers of organisms. Choose the index letter of the appropriate diagram to match each one of the examples shown in the table. Each letter may be used once, more than once or not at all.

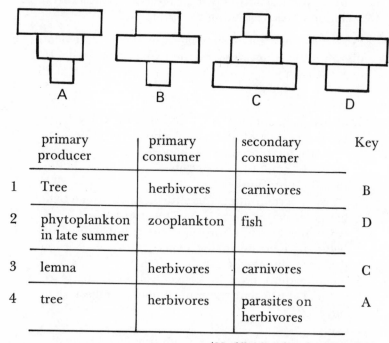

	primary producer	primary consumer	secondary consumer	Key
1	Tree	herbivores	carnivores	B
2	phytoplankton in late summer	zooplankton	fish	D
3	lemna	herbivores	carnivores	C
4	tree	herbivores	parasites on herbivores	A

(*Nuffield A-level trial 1967*)

Assertion-reason items are useful for testing complex thinking. Especial care is needed in setting them as the outcome of the process is expressed as a single code letter. Four examples are given, preceded by a general rubric on the answering code:

For each of the following sentences give

A if both assertion and reason are true statements and the reason is a correct explanation of the assertion;

B if both assertion and reason are true statements but the

reason is not a correct explanation of the assertion;
C if the assertion is true but the reason is a false statement;
D if the assertion is false, but the reason is a true statement;
E if both assertion and reason are false statements.

Key

1 Energy is needed for salt accumulation in vacuoles BECAUSE cell membranes are differentially permeable to large and small molecules.   B

2 All parasites have specialised organs of attachment BECAUSE they have to live permanently attached to their host.   E

3 Mendel concluded that genes were on chromosomes BECAUSE the random assortment of characters correlates with the behaviour of chromosomes during meiosis.   D

4 Transpiration often increases in the early morning, after the stomata are fully open BECAUSE the leaf temperature rises more quickly than the air temperature.   C

The chances of ambiguity in this type of item are great and they may penalise the more perceptive student who reads more into them than the examiner intended. It is best to validate them by asking students in a trial test to write out their reasons for their choice.

Structured questions requiring minimal writing

This method of assessment lies between the poles of the essay type question and objective testing. It can approach objective testing in precision and objectivity of marking and show an advantage over this in the information which they provide on the mental processes of candidates. They are particularly useful when designed as problems which test a range of skills which can be specified in a clearly defined manner, such as, translating and interpreting data, computing and handling results, analysis, formation of hypotheses and some aspects of practical work (except actual manipulation and direct observation). This possibility of relating to practical work makes this type of question the most suitable form of external reference test if a written paper is to be used for comparing the standards of different schools in their internally-assessed practical work. Structured problem questions have appeared in biology examination papers in increasing numbers during recent years, but their inclusion in mixed papers offering a choice of problems and continuous prose questions of equivalent length is not desirable. The 'all or nothing' effect of problem solving on a candidate's results can make an unwise choice disastrous. A candidate has to try a lengthy problem before he chooses it, often wasting valuable time. If specified behaviours have been included in the objectives of a test, then all students should be equally tested for them. This would indicate that problem questions should form a compulsory paper. The all or nothing effect indicates that they should be short. Of the various methods of testing, structured problems show good mark reliability and the best overall discrimination between candidates.

Short problems are a good means of testing for the transfer of skills from the context in which they have been practised in the course to new situations. Such course-independent questions are therefore more concerned with testing process than content. The following example (Nuffield Advanced-level biology examination 1969) is such a non-specific question as far as the course content was concerned and aims to test translation of numerical data into a graphical form in (a), knowledge of reasons for experimental procedures in (b), (c), (d) and (e), interpretation of results in (f) and formation of a hypothesis in (g).

An experiment was set up to compare the effects of different concentrations of IAA (auxin) on the growth of radicle sections and of coleoptile sections of oat seedlings. Five dishes, each containing twenty-five radicle sections 10 mm long and five dishes, each containing twenty-five coleoptile sections 10 mm long were set up and kept in the dark. The solutions used and the results when both radicle sections and coleoptile sections were removed and measured after 24 hours are shown on the table.

solution	average radicle section length	average coleoptile section length
1 2% sucrose only	11.5 mm	12.0 mm
2 2% sucrose + 10^{-5} ppm IAA	14.3 mm	12.5 mm
3 2% sucrose + 10^{-3} ppm IAA	12.2 mm	15.8 mm
4 2% sucrose + 1 ppm IAA	11.0 mm	16.7 mm
5 2% sucrose + 100 ppm IAA	11.0 mm	16.8 mm

(a) Plot as many of these results as possible on the grid ($\log_{10}$ IAA against length of section).

(b) Why were radicle and coleoptile sections protected from light throughout the experiment?

(c) Why is 2 per cent sucrose included in the cultures?

(d) Give two reasons why twenty-five sections of plant material were used in each dish.

(e) For what purpose was IAA omitted from one solution?

(f) From the results given in the table and plotted on your graph, what can you conclude about the effect of IAA on the growth of radicle and coleoptile sections?

(g) Suggest a simple hypothesis based on the results which would explain why radicles are positively geotropic and coleoptiles negatively geotropic. (Assume that concentrations of IAA within coleoptiles and radicles under natural conditions are within the range used in the experiments.)

This question was included in that part of the written examination which was used to moderate internally assessed class practical work.

Related to this type of testing is the setting of a passage of prose on a biological theme on which a series of short questions are asked. The particular use of this method is to test the ability of candidates to comprehend new information from reading and the ability to relate it to their background of biological knowledge. Such passages are used in the Nuffield Advanced-level biology examination.

The uses of assessment during a course
The extent to which formal examinations record achievement
in a course and predict future performance is being
questioned. Because formal examinations can only sample
limited parts of the work which has been done, they
introduce an element of chance. In spite of efforts to move
towards the testing of intellectual skills, there is still an
undue premium on memory and speed. Assessment of
students during a course of study introduces elements of
flexibility and can allow personal qualities to be displayed on
a wider front. On the other hand, the teacher also possesses
personal qualities and an element of subjectivity may well
affect his judgements. The long periods of contact between
teacher and students can introduce halo effects in which
grading of specific attributes is affected by overall impressions of the students. In spite of the possibility of some
unreliability on this account there has been an increasing
acceptance of some element of internal assessment by
students and this can be encouraged by class discussions on
the nature of the assessment procedures which are to be used.
The three main areas in which course assessment might be
applied are set written work, class practical work and
individual projects or dissertations.

Set written work
The main advantage of assessing this is that students have
access to resource material and memorising is not necessary.
Study in depth is encouraged and the student is less
influenced by the time element. As the work is produced
while the course is in progress, contacts with students can be
maintained for discussion and diagnosis. The use of this form
of assessment can be complementary to examinations which
include objective testing and problem solving. It also provides
more time for impression marking by two examiners than in
a final examination.

Class practical work
Assessment of practical work by formal examination might
be expected to provide impartial testing under controlled
conditions. The protracted nature of practical exercises,
however, allows few questions to be worked and this

increases the sampling error of the examination. Most practical examinations are concerned with assessing the results of observing and handling biological material and the outcome of a limited range of compact experiments. Very little can be done on assessing the performance of the necessary skills by actual observation of individuals. The variable nature of much biological material, the protracted and open-ended nature of many experiments makes them unsuitable for inclusion in examinations. The on-going nature of many investigations with the concurrent need for consultation and reference is better assessed by the teacher.

The dual tasks of simultaneously teaching and assessing are not easy and it may be felt that objectivity in particular is difficult to achieve. Precision can be increased by making use of a list of groups of skills which are to form elements of the assessment. Such a list (J. F. Eggleston 1968), related to Bloom's taxonomy of cognitive objectives, is given in outline.

A Knowledge of techniques and apparatus.
B Ability to apply techniques and to use apparatus.
C Ability to make appropriate observations and recordings.
D Ability to solve practical problems.

Such a list can be made more specific for a particular course. Decisions may then be made on the extent to which some of the objectives could also be assessed in written papers. An overlap between the two modes of assessment would allow the written paper, as part of an external examination, to be used as a valid moderating instrument for internally graded practical work.

The task of internal assessment should be kept as simple as possible. In the Nuffield Advanced-level biology course, which has fairly closely specified practical work, assessment is made under three divisions:

Procedure
Recording
Handling of results

This works well for practical exercises which are almost entirely designed as investigations used to provide evidence of biological principles. Each division is graded on a five-point scale with grade three as the class average. Teachers are encouraged to use the full range of grades wherever possible and intellectual outcomes are given more prominence than manual skills. Information for the assessment may be obtained from internally set practical tests, from selected class exercises or from a global judgement made over a period of time. Assessment of progress is a valuable diagnostic procedure, but to interpret it as a final grade is not easy. If assessment of attainment is required, it is unrealistic to use evidence from the early stages of a course. On the other hand deferring assessment to later stages presupposes that course work is homogenous and of a consistently progressive nature. This is not generally true of biology courses. Sampling of practical abilities must be related to the contrasting sections of a course.

Project work

Projects may be interpreted as critical surveys of available information on chosen aspects of biology, collection and utilisation of systematic observations of phenomena or the organisation of an experimental inquiry. It is in this latter type of project that a student is given most opportunities for personal initiative in determining the nature of his work and the means of carrying it out. Biology offers more opportunities than the other sciences for open-ended enquiries which do not require very specialised techniques and apparatus (Tricker and Dowdeswell 1970). The variability of organisms and the complexity of biological situations usually make it necessary to apply statistical processes to the interpretation of data. A student must plan for the end results of his project if its quality is to be acceptable. Guidance may be required in this so that a feasible and fruitful project employing practicable techniques is initiated. It is useful for the student to submit a written outline proposal for discussion with the teacher. During the working of the project, the teacher acts as a consultant and utilises his opportunities for assessment. The visible products of a project may include the working experiments and biological preparations, but the written report provides the most important evidence of achievement. This should be a unique communication as far as the student is concerned, showing evidence of the employment of a range of skills. It can be made available to a moderator.

The especial problem of assessing projects is their variety which makes comparison between them difficult. Objectives common to a range of projects can be used, provided that the nature of a project has been carefully explained to the students. The following list offers an outline which could be developed in more detail, according to the views which might be held on project work:

1 The selection of a defined problem
2 The design of experiments
3 The design and construction of apparatus
4 Manual skills and experimental techniques
5 Methods of analysis of results
6 Presentation and discussion of results

In an open-ended investigation, there can be no assurance of a positive outcome. Assessment should therefore be concerned more with process than with end results. There is no reason why several students should not contribute to a main investigation provided that individual efforts can be identified. This would be best organised as a cluster of problems, for each of which an individual could be responsible.

In helping students in difficulties, the teacher must distinguish between those due to shortcomings of the student and those due to the nature of the work. Records of such circumstances may have to be supplied to a moderator. Attempts at fine-graded quantification of skills are unrealistic. Usually not more than five grades are appropriate for each main objective, and three for the lesser ones (Eggleston and Kelly 1970). The criteria on which grades are to be awarded should be written down and might include attitudes such as initiative, a critical or perceptive attitude, interest, perseverance, conscientiousness, self dependence or cooperation. Any such inclusions can only be used if evidence on them can be obtained from observable behaviours. For example, a scale of five grades for 'design of experiment' might run as follows:

1 Much initiative shown in designing experiments. A perceptive attitude to the data required was shown.

2 Self dependence shown in designing procedures. Some discussion needed on planning details and data required.
3 Good choice of practical methods, but guidance needed on their details and in the selection of the data required.
4 Help required in choosing the most suitable methods from various alternatives and in the nature of the data required.
5 Inappropriate procedures proposed. Much guidance needed to start the project.

By attempting to work to defined specifications in various examination techniques, the teacher is more likely to classify students with some justice.

Summary
1 The objectives of assessment should be stated in behavioural terms.
2 Each method of assessment has its own advantages and disadvantages.
3 Methods should be selected according to the objectives (or the attributes of the students) which are to be assessed. A combination of assessment methods might provide more information for predicting future performance.
4 Teachers should be familiar with the variety of assessment methods which are being used. The assessment of work during a course should involve the teachers in schools as well as the external examiners.

References

J. Britton (1963) 'Experimental Marking of English Compositions Written by Fifteen-Year-Olds', *Educational Review* 16, 17 (November).
R. L. Ebel (1965) *Measuring Educational Achievement*, Prentice-Hall.
J. F. Eggleston and P. J. Kelly (1970) 'The Assessment of Project Work in A-level Biology. *Educational Research* 12, 2.
J. F. Eggleston and J. F. Kerr (eds) 1969 *Studies in Assessment*, English Universities Press.

J. J. Head (1965) 'Multiple Marking of an Essay Item in Experimental O-level Nuffield Biology Examinations', *Educational Review* 18, 3 (November).

W. D. Hedges (1966) *Testing and Evaluation for the Sciences*, Wadsworth.

R. E. Lister (1969) 'The Aims of Questions in A-level Biology Examinations', *School Science Review*, 172, March.

B. J. K. Tricker and W. H. Dowdeswell (1970) *Projects in Biological Science*, Penguin.

S. Wiseman (ed), (1961) *Examinations and English Education*, Manchester University Press.

Further Reading

J. M. Ashworth (1972) 'The Use of Multiple Choice Tests in Biological Teaching', *Journal of Biological Education*, vol. 6, no. 5 (October).

A. G. Callely and D. E. Hughes (1972) 'Analytical Questions in Exams', *Journal of Biological Education*, vol. 6, no. 5 (October).

Cuebs (1968) 'Testing and Evaluation in the Biological Sciences', *Publication no. 20, Washington DC*.

J. F. Eggleston (1968) *Problems in Quantitative Biology*, English University Press.

V. I. Ferguson (1971) *Continuous Assessment Tests in O-level Biology*, Heinemann.

E. J. Furst (1958) *Constructing Evaluation Instruments*, MacKay.

J. L. Harper (1972) 'Projects and their Assessment in Degree Examinations', *Journal of Biological Education*, vol. 6, no. 5, (October).

D. L. Nuttal and L. S. Skurnik (1971) *Examination and Item Analysis Manual*, National Foundation for Educational Research.

A. J. Rowe, (1972) 'Machine Marking of Multiple Choice Tests: A simple and Inexpensive System using a Desk-top Calculator; *Journal of Biological Education*, vol. 6, no. 1, (February).

Schools Council Examination Bulletins (1964) HMSO.

No 3 'The Certificate of Secondary Education: An Introduction to some Techniques of Examining'.

No 4 'The Certificate of Secondary Education: An Introduction to Objective-type Examinations'.

K. W. Thomas (1972) The Merits of Continuous Assessment and Formal Examinations in Practical Work, *Journal of Biological Education*, vol. 6, no. 5, (October).

R. W. Tyler (1931) 'A Generalised Technique for Constructing Achievement Tests', *Educational Research Bulletin* 10, Ohio State University.

R. Wood and L. Skurnik (1969) *Item Banking*, NFER.

Values and ideals

The need for change

COLIN STONEMAN

Why are innovations needed in education?
Innovations are not a *necessary* feature of education!
Changes in the content and method of teaching in schools
may be recommended with good reasons, some may take
place, but schools can continue to operate without innovation or reform because they are, to a large extent, immune
from external influences. Unlike industry, commerce, or
medicine, for example, where production, profit and the
relief of suffering are hard realities, schooling operates in a
more self-contained world in which its objectives, methods
and standards of success all exist *within* the system,
independent of the outside world. This may appear to be a
pejorative comment on one of society's benign institutions
but, if we are to consider innovation and change, it is
important to remember that schools do *not* have to adopt
new technology to avert financial disaster, or revise their
methods to reduce mortality, or face the crises commonly
found in other professions and institutions.

The notion of scholastic immunity to the changing world
helps to explain why so much advice from writers on
education has gone unheeded. None of the innovations
currently under discussion or on trial are really new; they can
all be traced to pronouncements of one or more philosophers
or educationists of the past. This is well illustrated by the
current trend towards greater participation by pupils in the
process of learning and the value of first-hand experience
instead of the teacher's word alone. This idea has been advocated by many eminent writers on education. For example:

Let all the lessons of young people take the form of doing
rather than talking; let them learn nothing from books
which they can learn from experience...

(Rousseau *Émile*)

> First-hand knowledge is the ultimate basis of intellectual life. To a large extent book-learning conveys second-hand information... The second-handedness of the learned world is the secret of its mediocrity...
>
> (Whitehead, *The Aims of Education*)

> To arrive at knowledge slowly by one's own experience is better than to learn by rote, in a hurry, facts that other people know and then, glutted with words, to lose one's own free, observant and inquisitive ability to study...
>
> (Pestalozzi, *Aphorisms*)

> The ordinary schoolroom with its rows of ugly desks... is all made 'for listening'—for simply studying lessons out of a book is only another kind of listening... There is very little place in the traditional schoolroom for the child to work...
>
> (Dewey, *The School and Society*)

These writers were concerned with education in general; it is not surprising that men of science who turned their attention to science teaching emphasised doing in place of listening to an even greater extent. T. H. Huxley and H. E. Armstrong were two of the most notable advocates for learning by direct experience and both made some impact on the educational thinking of their day (Armstrong, 1896). But in spite of many recommendations and arguments, classrooms at the beginning of the second half of the twentieth century were still filled with rows of desks and pupils listening too much and doing little for themselves. Though science learned in laboratories had a practical look about it, much of the activity was aimed at confirming the truth of the text book rather than true experiment. Reform and innovation were long overdue if the situation was judged by simple, common-sense criteria. Pupils, we might suppose, learn science by operating like a scientist on various aspects of the everyday world about them. This is not so. Even today, much science is taught by rote, in a hurry, as a mass of facts so that, glutted by words, the pupils lose their own free, observant and inquisitive ability to study.

Why are innovations needed in secondary schools?
It is neither true nor fair to say that all schools are unchanging. Education in this country has undergone much apparent change in its organisation and many new schemes for science teaching have been produced which are now in the schools. But even so, actual classroom procedures have remained remarkably resilient to change. It is in the primary sector that changes in practice have been most widespread and gained greatest publicity; in the education of the very young the principles of pupil participation and activity have been realised to the greatest extent. Project work, for example, is regarded as an innovation in secondary education, but has been an established method in primary schools for many years. This work encourages individuals or small groups to pursue the study of a topic of their own choosing. Apart from actual project work, many primary teachers encourage children to use the resources of the classroom and school on their own initiative, and many secondary teachers have been struck by the confidence and expertise in learning shown by eleven-plus beginners in secondary schools. Often initiative and motivation have been drained from these children by the traditional structure of secondary education—by the formal ethos, that peculiar blend of discipline, academic competition and frequent examinations which comprise the traditional school. Now the schools are under pressure not from remote philosophers and educationists, but from changes in school size, in pupil capability and expectation, and in examination structure. There is renewed interest in science which is not divided into specialist subjects. These, and other factors, are influencing the actual teaching procedure adopted in secondary schools and it is here that innovation is most needed.

Why are innovations needed in biology teaching?
Not only have teaching methods, in secondary education, been called in question but the *content* of some school subjects, have also been criticised, and biology is one of these. It entered the educational scene in England late, and rather diffidently, like a Cinderella beside its two well-established sisters, physics and chemistry. But, since the fifties, it has gained respectability and popularity; the reasons are not hard to find (Tracey, 1962).

Biology can (at least in theory) provide food for thought on all the great problems of our time. Through biology children can learn about people, over-population, genetics, sex and society, famine and pollution, drug addiction and conservation, IQ and brainwashing. In short, there is hardly a newspaper headline to be found that is not, directly or indirectly, a biological problem. Yet if we look at the syllabuses of most examination boards we find that great restraint has been exercised in the choice of subject matter deemed fit for pupil consumption. Furthermore, if we look at biology examination papers set for sixteen-year-old children we find few questions which hint that biology is a branch of science (see Table I).

Table I A comparison of mean percentages of questions set in biology O-level GCE by four prominent examinations boards, 1948–62, set out under four categories:

1 The direct acquisition of facts;
2 The interpretation of facts and the drawing of conclusions from experiments;
3 The application of scientific principles to new situations;
4 The designing and planning of experiments.

	Board A		Board B		Board C		Board D	
	48–59	60–62	48–59	60–62	48–49	60–62	48–59	60–62
1	78	84	90	91	84	89	85	85
2	2	1	0	0	0	1	0	7
3	1	1	0	0	1	4	4	3
4	19	13	10	9	15	7	12	6

(Crossland, 1963)

This curious state of affairs is not due to blindness or wickedness on the part of examiners. Syllabuses remain unchanged for periods of several years to be fair to teachers and pupils who try to grapple with them. The questions asked are, for the most part, factual, demanding only the recall of terms and principles, because such questions evoke a response from candidates which can be marked most easily and fairly. Unfortunately the educational system, in which examinations play a very large part, has produced a subject which is a compilation of descriptions and conclusions, not a method of solving problems which reflects the scientific work of modern biologists. (This criticism is not peculiar to science; students of history are seldom treated as young historians.)

Armstrong (1898) urged that all science teachers should treat their pupils as discoverers and used the term coined by Meikeljohn—the heuristic approach. It is as applicable to biology as it is to the physical sciences, but it did not flourish in schools because the ground covered by the approach was less than that of traditional methods; so students were said to be factually ignorant compared with their colleagues trained in a more orthodox manner. Such a statement raises fundamental questions about educational aims, but does not alter the fact that the heuristic approach to science teaching has not been popular.

Two major criticisms of biology teaching thus emerge. The work of pupils is largely passive listening, reading and memorising rather than active inquiry—an essentially scientific activity. The second criticism concerns subject matter, which is seen by many people as out-of-date in a world acutely aware of biological problems. A few of these problems are listed below, beside the main headings of an Ordinary-level GCE syllabus (see Table II).

What changes are currently being introduced?

Whereas at the turn of the century innovations and reforms were put forward by individual scientists, more recently such proposals have come from organised groups of teachers. Stimulated by professional bodies, the best known being the Association for Science Education, the Nuffield Foundation made large grants of money in the early sixties to aid the production of new teaching materials. Other projects which involve biology teaching have been sponsored in England, by the Schools Council, and in Scotland, by the Scottish Education Department (see Table III).

Though the Nuffield Foundation was responsible only for supplying financial aid, the style of science teaching described in the projects soon became known as the 'Nuffield

TABLE II

Biology syllabus: main subdivisions	Related, current problems commonly discussed but *not* encouraged by syllabus or examination questions
Food	Famine and world health
Nutrition	
Transport	Drug addiction
Food reserve and utilisation	
Growth	
Control and coordination	Learning, conditioning, brain-washing
Response to environment	Ecology, Conservation, Pollution
Reproduction	Population control
Asexual	
Sexual	Test-tube babies
Dispersal	
Heredity variation	Race
Chromosomes	Genetic engineering
Mendel's experiments	
Evolution	

approach'. This misled many 'lay' teachers and members of the public into thinking that Nuffield Science was something produced by a band of writers united by some bond of allegiance. This was not so; each project was run by independent organisers, yet there were common features running through the projects as a whole. There are publications too, which have developed from the projects. For example, the authors of *Biology by Inquiry* (Clarke *et al*, 1968) acknowledge direct links with the O-level Nuffield Biology Project. *Biology for the Individual* (Reid and Booth, 1971) is material 'intended to complement but not replace the Nuffield texts' and it is an adaptation of this material. So by borrowing, adapting and complementing, a nexus of materials has grown, all bearing certain family likenesses.

What are the characteristics of new science teaching projects? The broad aims of the five to thirteen Project (Ennever *et al*, 1969) are stated as 'developing an inquiring

TABLE III

Project with a biological content	Sponsor	Age range	Date of publication	Publishers
Science 5/13	Schools Council Nuffield Foundation Scottish Ed. Dept.	5–13	1972 on	Macdonald
Combined science	Nuffield Foundation	11–13	1970	Longman/Penguin
Scottish integrated science	Scottish Ed. Dept.	12–14	1971	Heinemann
Secondary science	Nuffield Foundation	13–16	1971	Longman/Penguin
Integrated science	Schools Council	13–16	1973/4	Longman/Penguin
O-level biology	Nuffield Foundation	11–16	1966/7	Longman/Penguin
A-level biology	Nuffield Foundation	16–18	1970	Penguin Education
16+ science	Nuffield Foundation	16+	—	—
Other Projects				
O-level chemistry	Nuffield Foundation	11–16	1966	Longman/Penguin
O-level physics	Nuffield Foundation	11–16	1966	Longman/Penguin
A-level chemistry	Nuffield Foundation	16–18	1970	Penguin Education
A-level physics	Nuffield Foundation	16–18	1972	Penguin Education
A-level physical science	Nuffield Foundation	16–18	1973 on	Penguin Education
Project technology	Schools Council	—	1966 (pilot scheme)	—

mind and a scientific approach to problems'. This introduces
innovation into science teaching because the aims suggest
that knowledge is something sought rather than received and
emphasises an *attitude* to be adopted by pupils. Orthodox or
traditional biology teaching has no formally stated aims but
appears to be concerned only with the acquisition of
knowledge, particularly knowledge of names and processes.

Knowledge

In the *Taxonomy of Educational Objectives* (Bloom *et al*
1956) the cognitive domain is well classified as knowledge of
facts and principles, comprehension, application, analysis,
synthesis and evaluation. Knowledge of fact and principle,
specifics and terminologies are sometimes referred to as the
'lower cognitives'. All the new science projects stress to a
greater or lesser extent, the *higher* cognitives—this means
greater emphasis on making judgements, putting ideas together, abstracting, and less effort to be spent on memorising
names. The ability to remember is just as important as it ever
was, but learning by rote, merely in order to recall, finds no
place in modern biology teaching.

Skills

The higher cognitives are associated with skills of a special
kind. It is increasingly recognised that children who have
undergone a course of study should be *better at* certain
things than children who have not. If the new project
materials function properly then students will practice the
skills of application, analysis, and evaluation, and get better
at them.

This country has been proud of the practical work done in
school science (including biology) compared with science
education on the continent, but the development of practical
skills has, in the past, been mostly a sixth-form matter. All
the new projects stress the need for pupils to find out some
things (not everything!) for themselves by direct experience.
In biology teaching this has meant, for example, the
introduction of microscopy and some dissection to the first
and second years of secondary school life—this is an
innovation.

Attitudes

Traditional biology teaching can be pictured (or caricatured) as the imparting of factual knowledge by a teacher to a class. The teacher explains as clearly as possible; the class listens and tries to understand and remember. Developing a scientific approach to problems is quite different and it requires changes in attitude on the part of all those involved.

The tasks confronting a student of traditional biology are circumscribed. There are certain areas of the subject which are already known by others and ready to be learned by the student who must direct his efforts to understanding, memorising and ultimately, exhibiting his knowledge. His attitude to teachers is normally one of respect, for they have already mastered the topic, and gratitude, if they can help him to understand it better and to learn faster. Attitudes to fellow students have little bearing on this kind of learning.

If the aims of biology teaching include the development of a scientific approach to problems, then these attitudes will not help. A student must be confronted with problems instead of statements and if a scientific approach is to be used, some of these must be genuine. That is to say, some questions must be open-ended with solutions which are unknown (or only partly known) to the teachers as well as the students. Faced with such problems a student and teacher can no longer be separated by a metaphorical counter across which one 'buys' and the other 'sells' information. There must be appropriate intellectual transactions going on in the classroom or laboratory and attitudes between one student and another are likely to be more important than hitherto, because problem solving is facilitated by the interaction of several minds rather than just one or two.

This kind of set up, which is commonplace when, for example, experimental projects in biology are being carried out, is far less secure for the teacher than the traditional situation. Though an experienced teacher has a distinct advantage, having studied longer and done more experimental work than the student, there are no guarantees that acceptable solutions will be found, or that the teacher will emerge as the wisest person in the class. The attitude of all concerned to the subject is also changed, for it can no longer be regarded as a set of circumscribed topics. Biology appears as a more life-like combination of 'attested fact and speculative theory' (part of a definition of science made by the Committee on Manpower Resources for Science and Technology, 1965). When speculation is experienced as the product of well-informed opinion and constructive criticism, then attitudes towards the subject are bound to change.

Assessment

If people strive towards some definite goal, it is only reasonable from time to time to see if they are succeeding or failing to achieve it. Schools generally, and in this country particularly, have been conscious, one might be tempted to say obsessed, with assessment and examination for the last hundred years. Until comparatively recently official examinations on school biology have been restricted to about a fifth of the pupil population, and the assessment has been done by the examination boards set up under the auspices of the universities. Their methods have been simple, overt and apparently impeccable, yet they have probably caused more harm to education than any other single factor.

The boards publish syllabuses listing a variety of biological topics considered worthy of study; an example might be feeding and the digestion of food. Examiners assess the candidates by setting short questions, such as, 'Describe the digestion of a piece of lean meat eaten by human beings. Why is digestion of food necessary?'. It would appear to be a test of ability in the description of a process taking place in certain organs in a certain sequence. The second part of the question calls for understanding of the significance of the process in the life of the human being. All seems fair until we look at the way in which such a question is likely to be marked. The availability of time and assistant examiners usually make it necessary to limit the process of marking to one of term and statement recognition. In this example, a mark or two for each stage, place of digestion, and for each enzyme involved, might be awarded. Likewise, key words and phrases in the second part of the answer are sought and marked. In other words, in terms of examination success, it pays to recall as much as possible and write as fast as possible. It does not pay to spend time selecting relevant from irrelevant information, choosing words carefully, making statements in the right order, or trying to understand the subject matter. Thus traditional examinations test a small part of a candidate's expertise (and many would say a relatively unimportant part), though they do test knowledge of facts.

Innovations in assessment have sprung from a desire to test more than this one part of a student's expertise, and to distinguish between his various capabilities within a given subject area, such as biology. Taking the same example as before, knowledge of the names of organs and enzymes can be tested; so can comprehension of digestion as an important physiological process; but not by the *same* instrument of assessment. Each kind of skill requires its own kind of instrument; if everything is tested by short, unstructured questions, the whole examination is reduced to a mere test of memory.

The variety of approaches which can be used to examine students following a course of study is quite well illustrated by the system used for Nuffield Advanced-level biological science (Kelly, 1970). The development of eight abilities are listed among the objectives of the course (Kelly and Dowdeswell, 1970) and the aims of *questions* in A-level biology examinations have been described in detail (Lister, 1969). The recall of factual information, for example, can be well-tested by objective tests, such as multiple choice and matching pairs. Ability to handle data and draw conclusions is assessed by structured questions which demand short, precise answers. Skill in expressing ideas in writing is measured by shorter questions and longer, essay-type answers. There is also a printed passage, taken from a text book, article or scientific paper, and questions are asked to test comprehension. The ability to carry out instructions in practical work, to observe, measure and record results is assessed at intervals throughout the course and there is an experimental project which is assessed. The project embodies a number of abilities which can be separately tested (Kelly and Lister, 1969; Eggleston and Kelly, 1970).

The assessment of the practical work, and part of the project, is left largely in the hands of teachers. This illustrates a second major 'innovation' which has appeared in recent years—the introduction of school-based assessment. This is not a new idea. The Norwood Committee of the Secondary Schools Examination Council (forerunner of the Schools Council) recommended in 1943 the replacement of the School Certificate by an internal examination controlled by teachers. Largely on the basis of this and the recommendations of the Beloe Report (1960), the Ministry of Education set in motion the machinery to establish a new examination system, the Certificate of Secondary Education which was introduced in 1965. It was new in that it attempted to assess pupils who were considered unable to pass GCE O-level and, more important, it was run by regional boards composed largely of teachers, and it could be school-based in operation.

Because a syllabus can be determined externally by a board or internally by a school, and because candidates can be tested from outside (by a board) or within (by the teachers), there must be four conceivable modes of examination. The CSE examinations covered three of these modes, as shown in Table IV.

TABLE IV

	External syllabus	*Internal syllabus*
External exam.	CSE Mode I GCE O-level	CSE Mode II
Internal exam.	GCE Pilot scheme	CSE Mode III

This is not the place to discusss the controversy over internal versus external examining. All agree that some measure of control and unification of standards has to be exercised by examining boards. This is done by 'moderating', a term which is used to cover a variety of procedures, each open to criticism, but beyond the scope of this chapter. It is important to note that these innovations in examining can profoundly affect the teaching of a subject like biology. A Mode III approach means that a teacher is free both to design his own course and participate in the assessment of his own pupils, subject to the approval of the board.

The GCE Boards have always been willing to consider syllabuses submitted by schools and to set special examinations (Hogg, 1968), but this has never been popular (probably because radical departures from the norm have not been accepted). They will also consider major modifications of their procedures as illustrated by a request, in 1968, by a small group of biology teachers to the Joint Matriculation Board (NUJMB), for what might be called a 'Mode IV' style (see Table IV). A pilot scheme was set up (Bennett *et al*, 1972) in which half the O-level assessment was done by tests, set by teachers and done in the school as part of the normal working arrangements, and half by a short written examination, set by the Board. In practice this means that a teacher in the scheme, who decides to deal with, say, digestion by means of experiments with enzymes, can devise tests to see how well pupils can handle apparatus, make observations, record results, etc. and submit the marks to the board. Neither he nor the class need to regard practical biology at O-level as an educational luxury devoid of value in the examination system. The teacher is prevented from becoming completely eccentric in his teaching by the other part of the examination which tests a wide selection of topics from the whole syllabus.

This touches on another area of controversy, whether a student should be examined on a few topics in depth, or more superficially over a broader sweep of the subject. The whole system in this country has been criticised for engendering too much specialisation too early in the lives of school-children (Jevons, 1969). This is more a matter of breadth of curriculum than a single subject but specialisation is a trend which can have effects in a subject like biology or in school science which are just as deleterious as they are on a larger scale. It has been proposed that the Advanced-level system should be replaced by one demanding the study of more subjects for at least one year in the sixth form (Vick, 1970) but whether or not changes will occur in this direction is uncertain.

One thing, however, is certain; official examinations have a profound effect on the kind of teaching and learning that take place in schools and two opposed views are held. One urges the improvement of the examination system so that what goes on in the classroom will, *pari passu*, be improved. The other view is that all examinations are intrinsically bad and that classroom practice will not improve until they disappear. This is not a far-fetched notion. Plans are under way for the replacement of GCE and CSE by a common examination system of some kind. Some people see this as a step towards the ultimate disappearance of a public, official examination for the sixteen-year-old child. They point to the fact that other countries manage well enough with fewer examinations for the school leaver and this age group.

The changes in assessment which have actually been introduced can be broadly described as analytical. They have stemmed from fundamental questions about the exact needs and functions of assessment. Why do we assess pupils, do we assess courses or teachers? Should we qualify children by a pass/fail check at the end of a course or should we discriminate further and award grades? What is the effect of grading on the child, the teacher and the course? More specifically, in the context of biology teaching, the need for well defined objectives has become apparent. In order to assess, one must first know what is being aimed at. Innovations in this field have appeared in a number of forms (such as a glut of objective test anthologies) but perhaps the best result is the diversification of test instruments to match a diversity of teaching objectives. We are now much clearer, as far as biology is concerned, about the kinds of knowledge and skills that pupils can acquire. This clarification has come, in part, from a search for valid methods of assessment.

The effects on children
The essential point of changing an educational method or system is that it shall benefit the students and this means that someone must decide what a student should become—what he should get better at—during his studies. Although one can make very heavy weather of educational objectives it is not a difficult task to list the salient and attractive features common to children. When young they tend to be inquisitive, imaginative, energetic, with considerable ability to memorise things which interest them. If we look at scientists it is not so very difficult to list a few of their most characteristic activities. Science education which attempts to combine the two lists, in some way, seems reasonable enough. Yet at university level 'studying science becomes stenography plus memorisation' (Jevons, 1969) and this pattern is too often

repeated at sixth-form level and throughout the schools. If some of the factual material is removed from a course and replaced, for example, by problem-solving exercises, then it seems fairly clear, from the limited research done, that the students get better at problem solving and know rather fewer facts than their predecessors. (Heaney, 1971). This may seem rather obvious but it is nevertheless important because such a statement can help to put new science teaching into some kind of perspective. Students by and large learn from the resources and materials put before them and the methods employed do not necessarily make them wise about other, related subjects, just because they are using inquiry methods. Harm has sometimes been done by making extravagant claims for new teaching methods (Martin, 1970).

Effects upon children, of changes in their normal learning methods, may be unpopular, at least in the short term. Discussions with sixth-formers doing Nuffield Advanced-level biology (Kelly, 1972) showed that they were interested and enthusiastic and prompted by the work to ask questions and generate ideas. But doubts were expressed about the adequacy of the factual content of the course, especially in relation to revision for examinations. This illustrates a very important aspect of newer biological studies—the lack of security through knowledge. The picture given by Jevons is one of happy if pointless activity—'There is much frantic scribbling in the lecture theatre, followed by evenings spent deciphering the results and committing them to memory...' When this activity is replaced by information—*seeking*, problem-solving and discussion of open-ended questions, much of the solid ground seems to slip from under the feet of those students who are intolerant of uncertainty.

Encouraging a critical attitude towards evidence, hypotheses and experiment is a good aim of science education. However, it can be brought to bear, not only on biological subject matter, but on the content of the course and on the teachers running it. One leader of a developing country has declared his rejection of new science education for this very reason—it stimulates a questioning of authority!

The effect of innovation in biology teaching on pupils will thus depend to some extent on their temperament and general ability, but they will also differ according to the type of teaching that the children are used to and how long they have been subjected to it. Much will also depend on the teacher. Whether or not changes in teaching resources and style will bring about much wider effects in the student remains to be seen. We are warned (Ministry of Education, 1969) that 'our future community will need not only more specialists... but also more citizens capable of imaginative and creative thinking within the context of science...'. It is difficult to see how we could ever know if we have achieved such an aim.

The effects on teachers

It has been mentioned that the effects of innovation on children depend partly on the previous experience of the child and this can amount to three to four years or five to six years for sixteen+ and eighteen+ students respectively. For teachers the period of previous experience must be longer because, even if they have only a year or two of teaching experience, they have about ten more years as a student to add to it. We have some evidence of the effects of *change* upon teachers but we have no clear picture of the effect of prolonged use of what are now called 'new' approaches.

For example, teachers taking a new set of teaching materials have found it necessary to learn new facts and get to know organisms previously unfamiliar to them. This has put them temporarily on a par with their students, no doubt with beneficial effects, but the situation soon reverts to the normal, in which the teacher knows and the student learns. Far more important is the change in *role* that is necessary if a teacher is to approach open-ended problems scientifically. The new position of the teacher can be illustrated diagramatically. The traditional function of a teacher of biology, and of practically anything else, was first to learn and understand the subject matter, then to digest and prepare it in a form for teaching, and then to impart the subject to his pupils (see diagram A). A teacher supervising a biological project can help in many ways, but not according to the traditional scheme of things. He cannot possibly prepare, digest and teach, for this would mean doing the student's work for him (see B). Not all biology teaching is done by projects, but if the teacher is using resources and materials published or manufactured outside the school, then his role still differs from A; he becomes more of an *organiser* than a person imparting knowledge (see C).

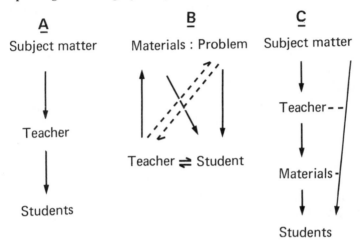

If we add to the list of novelties, programmed learning, individualised learning and team teaching, the need for organising ability is very great indeed. A teacher must know what the various aids, materials and resources are supposed to *do* for his pupils rather than what they are *about*. This organising role does not suit teachers who feel confident only when they are familiar with all the knowledge and skills the students are supposed to acquire. It is again partly a question of security and partly to do with the notion of an ideal teacher. New materials not only suggest a new role, they demand a change in the teacher's physical (as opposed to metaphorical) position in the classroom or laboratory, and this can bring about a change in the relationship between those who teach and those who learn.

Perhaps this is the most profound and important effect that changes in teaching method can bring to biology, or any other subject. Much has been said and written about teacher-pupil relationships. It is very difficult to attempt any change when the subject matter, the teaching resources and the examination system all conspire to drive teachers into

their territory between the blackboard and the demonstration bench and pupils behind their desks. Many still believe
that teaching is really proclaiming something verbally and
that other styles are, in some way, an abdication. But, as was
mentioned earlier, school environments in which the declam-
atory mode is possible are fast shrinking.

The effects on schools

The schools (together with the bodies who finance them)
have to provide an appropriate environment if biology is to
be taught. Proposals for new styles and schemes have often
met with objections, based on lack of funds for apparatus
and equipment, lack of room to allow sufficient bench space
for large classes doing practical work and, occasionally, lack
of time (the number of periods devoted to biology). One
thing that has come out of a decade of science curriculum
studies in this country is the considerable inequality of
resources between otherwise similar schools. It is a question
of norms—are some of the new schemes excessive in their
demands for resources or is the old system unduly mean
towards teachers and students? So one effect of proposed
innovation is that the status quo is questioned—some kind of
yardstick can be applied to school biology and science
departments.

A more precise effect appears in the planning of laboratories. In the old days one knew what the subject was and
could install benches and apparatus to meet its demands.
With the call for more scientific work in the form of
individual projects, or group research, it is impossible to
describe the kind of procedures which may be necessary from
one year to the next. Laboratories must therefore be made as
flexible as possible to accommodate changing activities.

Effects on the curriculum

A restless and changing subject, which is concerned with
discovery, is unlikely to remain within tidy boundaries. When
does biology become biochemistry? What does one do about
a topic or problem which turns out to be largely a matter of
physics? In the 1930s general science was in vogue and was
recommended to be taught in all secondary schools, but it
never gained universal acclaim, and slumped almost to

nothing by the end of the 1950s (James, 1957). Now we see the same spirit expressed in the production of several combined or 'integrated' science projects. This renewed enthusiasm is partly an effect of the new style: there are other reasons for it, including availability of qualified science staff. This time the cause may be more successful though some of the old constraints still apply. One constraint was the desire of each teacher to stick to his own specialist science, but the growth of team teaching and individualised learning may, in their separate ways, compensate for any such lack of flexibility.

Another form of re-alignment, due to rethinking and innovation, is to make two sub-sciences out of three established subjects—physical and biological science. This recognises certain overlaps in conventional physics and chemistry, a certain amount of unduly academic material which might well be cut out of modern syllabuses, coupled with the need for a somewhat enlarged and modernised form of biology. So far, new schemes of work have been limited to biology 'pure' or combined with physics and chemistry, but it must be remembered that rural science (rural studies) and human biology are specialised aspects of the same science. The future of these in the curriculum is uncertain.

It has been said that subject matter does not impose its own unique teaching style (Jevons, 1969). This may be true but it can *influence* style, and the reverse is certainly not true, teaching style can *impose* limits on subject matter. If biology teaching continues to develop towards more inquiry and experiment, then its subject matter will be affected. It may be that a good teacher can teach any topic in any style, but some things will be a lot easier than others. For example, it is far more difficult to encourage hypothesis-making through observation of prepared sections of mammalian tissue than, say, pulse rate or analyses of exhaled air. Plant anatomy does not lend itself to project work quite so readily as plant physiology. In the past, topics for the syllabus have been chosen by biology committees of examining boards, as a set of things considered worthy of study. One can foresee a change of emphasis towards those things which produce good teaching material, things which, for one reason or another, are found to have an educational value.

The future

Since 1963 there has been a decade so filled with science curriculum development that a biologist might be tempted to regard it as an intense burst of mutation. It might not be a bad thing if this were followed by a quieter period of 're-assortment' and 're-combination', in which the various new educational items could be sorted, adjusted and modified into recognisable patterns, without necessarily fixing them for all time, as a kind of new orthodoxy.

During such a 'resting' period, at least two things are needed by the teaching profession. One is a means of easy communication, so that the experiences of science teachers can be shared and compared. The other is research and evaluation of new methods and materials, which is *not* the same thing as mere communication of results. Educational research has suffered in the past from lack of stature—it has often been too sporadic, idiosyncratic and done on too small a scale. Education is full of unwarranted assumptions, mixed with benign optimism—we need to *know* what really happens when we teach children in a certain way. Evaluation, like examinations, should involve the teachers, and the results should be an aid to their work, not just a gain for the researcher.

Finally, we must hope that in some way schools can lose their immunity from the outside world, which was mentioned at the beginning of the chapter. Biological science is about the real, external world, and it should not be too difficult to devise means of preventing an inward-looking attitude developing in science education. We must avoid this attitude, so prevalent in the past, which led to a self-perpetuating, self-evaluating and self-centred kind of subject.

References

H. E. Armstrong (1896) 'How Science must be studied to be useful', *The Technical World*.

H. E. Armstrong (1898) 'The Heuristic Method of Teaching', *Board of Education, Special Reports on Educational Subjects*, vol II.

D. Bennett *et al*, (1972) 'School-based assessment as part of O-level Biology GCE', *School Science Review*, 53.

B. S. Bloom *et al*, (1956) *Taxonomy of educational objectives. Handbook l, Cognitive domain*, Longmans.

R. A. Clarke *et al*, (1968) *Biology by Inquiry*, Books I—III, Heinemann Educational Books.

R. W. Crossland (1963) An address to the Science Masters Association University of Manchester, (4 January 1963).

J. F. Eggleston and P. J. Kelly (1970) 'The assessment of attainment in project work for A-level Biology', *Journal of Educational Research*, 12.

L. Ennever (1969) *With objectives in mind – an introduction to Science 5/13*, Schools Council Publications, Macdonald.

M. E. Hogg (1968) 'Nuffield O-level Biology Project. Examinations and Teachers in Training', *Journal of Biological Education* 2.

J. O. Head (1972) 'A report on the progress of science undergraduates who had followed Nuffield Advanced level science courses in schools'. *The Times Educational Supplement*, 2 June 1972.

S. Heaney (1971) 'The effects of three teaching methods on the ability of young pupils to solve problems in biology', *Journal of Biological Education*, 5.

W. S. James (1957) 'The slump in Science teaching', *New Scientist*, 19 September 1957.

F. R. Jevons (1969) *The Teaching of Science*, Unwin University Books.

P. J. Kelly, (1972) 'Evaluation Studies of the Nuffield A-level Biology Trials: 5', *Journal of Biological Education*, 6.

P. J. Kelly and W. H. Dowdeswell (1970). 'Nuffield A-level Biological Science Project', *Journal of Biological Education*, 4.

P. J. Kelly and R. E. Lister (1969) 'Assessing practical ability in Nuffield A-level Biology', in *Studies in Assessment* (J. F. Eggleston and J. F. Kerr eds). Universities Press.

R. E. Lister (1969) 'The aims of questions in A-level biology examinations'. *School Science Review*, 50.

P. G. Martin, (1970) 'The case against Nuffield O-level biology' *School Science Review*, 51.

Ministry of Education, (1960) *Science in Secondary Schools*, Pamphlet 38, HMSO.

D. Reid and P. Booth (1971) *Biology for the Individual – Introduction*, Heinemann Educational Books.

G. W. Tracey (1962) 'Biology – its struggle for recognition', *School Science Review*, 43.

F. A. Vick (1970) 'The Curriculum and Examinations in the Sixth Form', Report of Conference convened by the Committee of Vice Chancellors and Principals, 19 Feburary 1970.

Further Reading

Association for Science Education, AMA and AAM, (1970) 3rd edition, *The Teaching of Science in Secondary Schools*, John Murray.

J. S. R. Goodland (1973) *Science for Non-Scientists*, Oxford University Press.

F. R. Jevons, (1969) *The Teaching of Science*, Allen and Unwin.

R. A. R. Tricker (1967) *The Contribution of Science to Education*, Mills and Boon.

Flowers and human ecology

HUGH ILTIS

A love of diversity

Why do I get an irrepressible urge to defend what I love, the beauty and diversity of nature, and, especially the disarming loveliness of its flowers?

One of my earliest recollections is joyfully picking huge and wildly unorganised bouquets of flowers on a Moravian mountain meadow, scabiosas, bluebells and daisies, and then lying on my back in a 'nest' surrounded by tall, tall grass watching the bees and the clouds. Ever since then, I have been an addicted botanist, 'half-plant', as an old friend of mine used to describe me.

My family was always interested in natural history and that tradition gave me much botanical stimulation and often a rather pointed direction to become a botanist (sometimes whether I liked it or not!). Besides being a professional botanist, and an advocate of adult education, my father was also a *preservationist* already in the 1920s; in a simplistic way, he tried to preserve an acre here and there for its rare flora. Way ahead of his time, he was singularly unsuccessful. Then hardly anybody thought preservation worth while, not even the Socialists, among whose ranks he was active, for the archaic creed that the 'people's' needs must always come first, no matter what, was then as now an often ill-applied battle cry against injustice.

In any case, by the time I went to the University of Tennessee in Knoxville I was well prepared to appreciate Professor Jack Sharp, a former student of the still active environmental pioneer Paul Sears. In 1945, Sharp botanically explored the Mexican Sierra Madre and returned with grim tales of horrendous erosion, increasing overpopulation, and outright destruction of forests. Despairing, but never silent, he kept hammering at the issues, at the blind insanity of both hungry and greedy men in a world spinning out of control.

But the nearby Great Smoky Mountains National Park offered an alternative lesson—nature can be preserved for its own sake, remain unharvested, and still serve man most genuinely. It was obvious we needed to preserve more national parks, more wild land.

Four years at the Missouri Botanical Garden in St Louis provided my professional training. There were trips to the Ozarks with botanist Julian Steyermark whose early voice argued for preservation of rare habitats. There were classes with Edgar Anderson who, while a fabulously gifted botany teacher, never took any real interest in preservation. This kept bothering me, for his lack of concern was the rule among the botanists there, rather than the exception. It seemed that many, perhaps most, and even the most excellent, professional biologists felt no social responsibilities towards nature, or even towards the very organisms they busily studied. I had one or two teachers who could not care less and said so. The less they cared, the more I wondered what sense there was in revising a genus of plants when it would soon become extinct. And what a pity not to save that which is so beautiful and gives so much joy.

Like all graduate students at the MBG, I took a trip to the tropics with a fellow student. We went to Costa Rica, where we wondered at the tropical flora in more ways than one. Behind the Instituto de Agricola Inter-americana in Turrialba, along the Rio Reventazon, there was a small protected patch of virgin tropical rain forest. It contained what still is probably the largest population in the world of a most remarkable giant tree, *Oreomunnea (Engelhardia) pterocarpa,* of the walnut family, representing a group of 'living fossils' with an excellent seventy million-year-long paleontological record. Of this species, there exist only a few hundred trees in the world, and about fifty were here in this little patch of timber, this sylvan cathedral, their crowns reaching to the sky. What a marvel! Yet several of the more 'practical', forestry-orientated botanists at Turrialba thought this forest, and with it these rare trees, ought to be cut. Thus, not only were some botanists not preservationists, some were even professional environmental rapists. That thought alone still makes me sweat! How senseless, how blind can men be? Surely, if only fifty of anything are left, they demand protection. Thus, slowly, slowly but surely, my childhood love and aesthetic awareness of nature matured into a passionate desire to save it.

The preservation of nature as a passionate concern

When I came to the University of Wisconsin in 1955, I was flung head-on into an old and well-established preservation tradition, a new way of thinking about the landscape and its preservation. There I found organisations called 'The Friends of the Native Landscape' and 'Citizens Natural Resource Association', as well as many concerned new friends and many new books, such as Aldo Leopold's *A Sand County Almanac*. They taught me to open emotional and intellectual doors previously kept shut by timid 'hang-ups' about the seeming sentimentalities of nature preservation, by feelings of embarrassment, which our carefully orchestrated culture, based on nature's destruction, ensures we all have in abundance. In Wisconsin, a state board for preservation of scientific areas administered choice scientific areas preserved solely for teaching and research. Its members, including the famous ecologist J. T. Curtis, supported a new concept: that the state, the body politic, is ultimately responsible for the preservation of nature; that to preserve flowers you have first to preserve and manage their total habitat; and that, if you do not thus ritualise preservation and sanctify it into law, there can be no hope for fish, fowl, or flower!

Not only the state, the sum total of all its citizens, but all citizens themselves must be personally involved in the preservation of nature. But that brought up some very difficult questions. Who will pay for this? What if a poor farmer owns a magnificent virgin oak-maple-walnut forest on part of his land? Is it *his* duty to save it, voluntarily and without recompense, for university research? Is it *his* responsibility to keep this ecosystem intact, and as a consequence deprive his family and sacrifice his profit for the long-range good of posterity, or for some well-to-do city folk to enjoy the wild flowers? What had posterity, or the wild flower lovers, ever done for the farmer? Clearly, the fate of nature cannot be left to the farmer's altruism. There are some who would perhaps wish it to be so. Yet this is quite unrealistic. Rather, preservation of nature and of diversity must be considered the duty and responsibility of the state because it benefits all its citizens. It must, therefore, be paid out of public funds and administered as a governmental function; its preservation cannot be left to chance or to the goodwill of a farmer. If the state is unwilling, or as yet unenlightened enough, to purchase such forests or prairies, then private citizens should organise and buy the land.

Nature preservation was often presented with promises of economic reward as the main rationale: the foods yet to be discovered, the drugs yet to be found, the lovely shrub or bulb yet to be brought into cultivation, the breeding stock to be saved—nothing useful to us or our children should be allowed to become extinct. The utility of nature preservation, especially of scientific areas, was a main theme of many a lecture (Iltis 1959). Despite its flaws, this still makes much sense, and, to many people, remains the only valid argument.

However, basing preservation on such purely utilitarian views left out any genuinely natural reasons. Gradually, some rather subtle but very basic questions kept intruding—what if people simply required unspoiled nature with its patterns and diversities to be happy, or simply to maintain their health? Did we not evolve within nature? Are we not adapted to it?

In 1962, the environmental crisis broke, suddenly, with *Silent Spring*. Even while Rachel Carson was publicly ridiculed as a sentimental old lady, and the ecological concern over DDT as silly, the subtle necrotic effects of pollution became only too apparent. It began to look as if DDT and mercury compounds might inherit what was left over after cow and plough. In the meanwhile, from colleges of agriculture to big chemical corporations, the organised opposition to any sensible approach was making sure that the public would remain deluded and misinformed. Profit was profit—to hell with nature.

More primitive cultures—more natural ways—
more tranquil humans
During this time, two trips to Latin America stimulated a major turning point in my thinking. One, in 1960, took us to Mexico to search for *a natural and quickly degraded*

insecticide, the 'Sabadillo' (*Schoenocaulon*) of the lily family, and another, in 1962, to Peru to study the evolution and taxonomy of wild and weedy potatoes. Travelling through primitive villages and ancient towns we saw Indians in a new light. In the high Andes or Sierras especially, the natives lived *with* the land, in it and from it, and, poor as they were, only rarely deliberately destroyed it. Perhaps this was merely natural, because they did not have tractors or other technological weapons and could scarcely be violent with it. In any case, uneducated, even illiterate as they were, the Indian men and women knew a great deal about both wild and cultivated plants, where they grew best, what they were good for, and which they loved best for taste, colour or form. They knew by common name hundreds of wild species of plants; and each of even the most subtly distinguished 'criollo' varieties of potatoes, for example, had its own common name. The phrases *es bonito* or *es linda*, often applied to a flower or to a brilliantly coloured ear of corn, transferred to us their gentle appreciation of natural beauty.

They were also gentle and undemanding of their children as well. I watched women breast-feed their young children as evolution had meant that relationship to function, on roadsides, on buses, in a calmness and naturalness quite rare (at best) in hectic suburban America. There were no plastic bottles with canned formulas here, only an innate, quiet understanding of the nurselings' needs and the undemanding relaxed love of their mothers (cf. Montagu 1972). These babies were suckled for two or even three years, and if this causes surprise, it is well to remember that our closest kin, the chimpanzee, may suckle its young to the age of four!

The contrast between that blessed peace and our frantic turmoil kept obsessing my brain. In the villages, even older babies were always carried, usually by their mothers, or by older brothers or sisters. They hardly ever cried, which one could hardly say for offspring in urban America. When they were hungry, they fed, when sleepy, they slept. Some questions came back again and again: if, in their natural way, these mothers were fulfilling their young children's natural, innate demands, was that the ultimate reason for the tranquility of the adults? If babies have to have mothers to nurse from, and mothers to be loved by in order to be happy

and tranquil, may not humanity similarly require nature to
live in, and nature to interact with, to be happy and tranquil?
For us, after all, nature should know best: did we not evolve
to fit her ecological requirements; or better, were not our
ecological requirements determined by her? These analogies
may be somewhat misleading. Nature does not always love
us, by any means. Yet they do allow us to ask a very basic
question: what is it, in nature, that man needs? That makes
him happy?

Is there a basic optimum human environment?

The insight hit hard one day, that love for and involvement
with nature, as a human trait nearly universal among people
both primitive and civilised, may in part be genetically
determined; that human needs for natural diversity and
beauty, like human needs for love, must be inherent (Iltis,
1966). Man's love for natural colours, patterns, and har-
monies, his preference for forest-grassland ecotones which he
recreates wherever he settles, even in drastically different
landscapes, must be the result (at least to a very large degree)
of Darwinian natural selection through eons of mammalian
and anthropoid evolutionary time and even through the ten
millennia of human agricultural history. Our eyes and ears,
noses, brains, and bodies have all been shaped by nature.
Would it not then be incredible indeed, if savannas and forest
groves, flowers and animals, the multiplicity of environmental components to which our bodies were originally
shaped, were not, at the very least, still important to us?
Would not such a concept of 'nature' be a major part of what
might be called a basic optimum human environment?

One could elaborate this insight in a thousand obvious as
well as more subtle ways. Yes, air to breathe, water to drink,
plants and animals to hunt and eat, sounds to hear, images to
see, textures to touch, land to wander on, people to love.
But, moreover, does not spring, with sprouting meadows and
flowering trees, stimulate joy? Do not then lovers go courting
in the country, flowers in hand, sensitive as never before to
the 'feel' of the awakening, growing land, to every odour of
blossom or lover, intoxicated with the dark, demonic urges to
dance, to touch, to mate close to the earth? And do not boys
and girls fish, men hunt and fight, both men and women

garden? Do we not, if we have money, build swimming pools or greenhouses, or second homes in the mountains? Do we not stream out of our cities by the million to camp in nature every spring and summer, a weekend exodus of human lemmings? Do we not, if we have a choice, keep pet dogs, cats, or fish, *cyclamen*, *primula*, or *begonia*, and humidifiers and flickering fireplaces in our miserably cold northern winters as ethnological fragments, almost symbolic bits, of that ancestral ecosystem which shaped our physical needs and subconscious desires? Could this also be the reason why some of us get seduced by surrogate plastic flowers, or even plastic trees (Krieger, 1973)? Is it to satisfy such basic innate needs that we wish to keep in touch with nature?

If there is then a biologically valid concept of a basic optimum environment for mankind, would it not have to include, because of the innate needs of the original naked apes and ape men, not only mothers for babies, but much *real* nature for children and adults? Is not this optimum human environment, modified somewhat, to be sure, during hominoid evolution, definable in its essence as a compromise containing maximum contact with nature without giving up all the advantages of culture and civilisation? And would not human needs for this nature-contact be of different intensity and quality for different ontogenetic stages: water, mud and love for babies, shrubs to hide behind and protected excitement for six-year-olds, groups to defend, fields to dig in and grass to sleep on for adolescents and adults? How long since we left the savannas? the tribes? the villages? Surely not long enough for all of us to have lost all need for their natural settings?

Here, finally, was an argument for nature preservation free of purely utilitarian considerations: not just clean air because polluted air gives cancer; not just pure water because polluted water kills the fish we might like to catch; not just saving plants or ducks because they could be 'useful' or 'edible'; but preservation of the natural ecosystem to give body and soul a chance to function in the way they were selected to function in their original phylogenetic home. The ultimate argument for nature preservation, as well as for landscape architecture or urban planning, would then rest squarely on evolutionary principles.

The optimum environment of any organism is the one in which it evolved, whether man, grasshopper, water lily or amoeba. This is given—it has to be, and no one has to prove it. On the other hand, most of us city dwellers live in a human environment which is but a few thousand, and to most human genotypes, only about a hundred years old. In 1810, only about 8 per cent of the world population lived in towns; today it approaches 50 per cent. Today, many of these people live in metropolises which, in many crucial parameters, are *not* optimum environments by any stretch of the imagination. And why not? Because, in them, we are like wild colts which evolved to roam the steppes but are tethered in a dark cave. Unlike rapid cultural evolution, physiological and physical evolution is an immensely slow process in which a hundred and sixty years (or eight sexual generations) will produce no, or at best only the most minute and imperceptible, changes in our basic adaptations. True, we must have slowly evolved into more docile creatures during the 10,000 years of predominantly agricultural life, and today, especially since the industrial-urban revolution, we are still evolving, perhaps even rapidly, in response to urban selection pressures, but whither and to what final fate no one, of course, can say. While perhaps in 20,000 years we may evolve into a half-blind, half-deaf *Homo sapiens* subspecies, *post-sapiens*, composed of environmental morons programmed to tolerate pollution, noise and ugliness, and perhaps allergic to the confusing diversity of natural conditions, today we are still part of *Homo s. sapiens*, a subspecies with strong genetic needs for the natural environment, whose members, depending on individual variability, try at every turn to find nature surrogates, from air-conditioned humidified domiciles, to aquaria with fish, to plastic trees designed by well-meaning landscape architects.

Clearly, the finest human and artistic attributes of even the most civilised man and culture have their roots in nature. Judging from the intensity of man's feelings and his search for substitutes we may yet preserve these roots and their substrate as the basis of his humanity. The substitutes we find for our original habitats are legion, and many *are* important and satisfying. Clearly, we cannot all go back to the African savannas, for there are too many of us! And even

if we could we would not want to—for most of us may have
lost the ability to live and hunt like cavemen. Yet if,
depending on age, climate, and season, we wish to experience
a highly optimal physiological-psychological environment, we
must retain the option of replicating the environment of our
ancestors thousands and tens of thousands of years ago, at
least now and then. And that means a return to biotic
diversity and adventure, reaching from the simplicity of a
colourful flower garden all the way to the unmanipulated,
undepleted complexity of a wilderness. And, above all, it
means the opportunity for intimate sensory contact with
natural beauty, both cultivated and wild. We do need
well-kept colourful gardens, as we do fields of grain and fibre.
But to me at least, in terms of sensory contact, these are only
special substitutes for wild land, with crucial yet limited
functions and usually with low diversity.

Thus, today, for the sake of diversity, we have ever more
urgent reasons to preserve the many, many species of plants
and animals which cannot be grown in gardens or maintained
and reproduced in zoos. Except for a few domesticated
species, this can *only* be done by protecting their original
wild habitat, by preserving the very special ecosystem to
which each species is adapted and of which it is a part. (Even
for man, despite his immense adaptability, only time will tell
how strictly that great principle holds true for him as well.)
This is why we need so many nature preserves, whether small
scientific areas and county parks or giant wilderness areas of
immense size, because there are so many species and
ecosystems to be protected! This is why we need international agreements on animals and ecosystems, subsidies and
concerted efforts for the preservation of the tropics and the
oceans on a scale as yet hard to imagine. And for the
preservation of many habitats even tomorrow may be too
late. Thus, and finally, this is why every biology teacher must
be passionately committed to the establishment and protection of biotic preserves of whatever size. Not only
membership in preservation societies is required of the
teacher, both to help financially and personally in the task
and to give every student a shining example of genuine
commitment, but the biology teacher must transfer to the
student the respect, the love, and the value such preserved

and undeveloped areas have to society. He must in fact teach the student how to use them and protect them.

In arguing for preservation of nature, the diversity of wilderness is not the only concern today. We need to preserve *human* ecosystems showing high diversity, stability and natural beauty. Thus, apple or peach orchards, vegetable plots and fields of specialty crops such as artichokes, or small low-yielding but diverse farms on the outskirts of cities are all agricultural communities, essentially ecotonal in character, combining forest and grassland into a kind of anthropogenic savanna of great aesthetic and ecological value for many people. Such small farms with many kinds of crops and groves of trees are quite stable, unlike the monocultural systems of Montana wheat or Iowa corn. The picturesque landscapes of Normandy, or of my native Czechoslovakia, the alternating farms and forests of southern Ohio, the mixed farming area of the Ontario Peninsula, and the unique combination of farming, market gardening, woodland and lowland in Kent and Sussex also deserve our concern. Now much of such land is being deliberately subdivided in the name of growth and profit or developed and destroyed in the name of progress in a global gesture of supreme carelessness. Biology teachers and planners, especially, need to explain to their students the elusive meaning of this loss to man and offer them viable alternatives: statutory limits to the growth of suburbia, the benefits of good zoning, with changes in taxation structure to allow farming next to apartment houses, and above all a deliberate policy of restraint on growth and on making money out of the selling of land.

What new research would help humanity
construct a more livable world?
The general malaise of mankind, especially of Western civilisation, would make it appear that urban humanity has become the unhappy involuntary guinea-pig of a giant experiment in sensory deprivation (coupled to lopsided sensory overstimulation), in which the crowded cities are our cages, and the inexorably cumulative acceleration of technology and the loss of nature the experimental variables. Nobody really is exclusively to blame—not technology *per se*, not capitalism, nor overpopulation. But the syndrome is such

that nobody can get out of participating in this experiment (except temporarily, perhaps, the young drop-outs that pack a tent and go back to the woods). True, while we have reached much higher standards of living and a longer life expectancy, there are other basic measures of an optimum environment, measures which would make it appear that we are reaping the synergistic effects of this sensory deprivation-overstimulation syndrome on a gigantic scale. The sea of broken homes; the epidemic of first unwanted, then unloved, then battered children; the ill-disguised anger of many childless and husbandless women seeking surrogate solutions for their biologically unfulfilled lives from *Cosmopolitan*; the blandness of purposeless males seeking salvation in *Playboy* philosophies; and the increasing alienation of young and old alike, both from society and from nature, all speak of unpredicted, unwanted consequences of our high culture, if not of an inescapable flood-tide of social pathology, and certainly not of an optimum environment to which humanity is genetically adapted.

Explanation for this gradual collapse of social structure and biological intercoherence (of man to nature, including other men) are legion, and include crowding, pollution, and the unbearable continuous roaring noises of the large city. Significantly, recent research has shown mental disease to be twice as high in New York City than in rural northern Illinois, a not unexpected conclusion. Yet why should this be? If we even cursorily analyse the difference between these two habitats of man, New York City and northern rural Illinois, we find that in addition to the *lack* of noise, pollution, and crowding, there is the *presence* of beautiful green plants, which to us mammals have always been either 'neutral' and unthreatening or even inspiring, as well as to a variety of animals, all combined into an infinitely varied landscape. Could it then be that the stimuli of non-human living diversity makes the difference between sanity and madness? If so, is it not about time to find out why? Is it not finally time for experimental psychologists, human etho-logists, and landscape architects or urban planners to get together and design experiments, scientifically indisputable and clinically impeccable, which would give us insights into how the brain of the human animal reacts to natural diversity and quality (beauty), to technological sterility, and what exactly happens when we look at the symmetrical pattern of a leaf or smell the odour of a brilliant flower? With modern EEG equipment and a little imagination, it would be rather easy to design experiments, using a diversity of groups of different ages and with diverse social and ecological backgrounds, comparing children with adults, farm youngsters with high rise apartment dwellers, the emotionally healthy and the mentally sick. Hardly anything is known about the way we neurologically relate to our surroundings. Could it be, perhaps, that we don't want to know? That if the truth were out, many of the powers that be would then be forced to stop making a desert out of the green earth, which would cut into profits, among other things, and rising GNPs?

The truth of the matter is that hardly anyone has bothered to think about the connection between the alarm of the 'preservationists', the social pathology of the cities, and the overwhelming ignorance regarding human needs in general and especially for the aesthetics of nature, a subject generally reserved for the abiological disciplines of the humanities. It is thus interesting to read at the conclusion of *Reflections on Environmental Education* (Stillman, 1972) a very stimulating, if often rather derogatory and ill-informed attack, on 'prophets of doom' (such as this author), the following perceptive admission:

> I have been able to find nothing in the environmental literature to help me understand what it is in natural environments that inspires human imagination, response, and excitement. Whatever this is, it should be added to the tool-bag of sensitive and imaginative teachers.

Indeed it should! For the 'whatever' Stillman is talking about is our innate response to the natural environment, a response selected (like us) over many millions of years. It is a natural human response to our original basic optimum environment, to the world of nature. There have indeed been no efforts made whatever to establish in a clinical and scientific way just *how* this interrelationship functions. Yet, the above insight is not all that ingenious. Wise men since time immemorial have subjectively sensed the important relationship of human happiness to natural conditions and incorporated it into ethics and religion. And artists have had it in their bones since the beginnings of recorded cultural time in the torchlit caves of Lascaux. Most art itself is but a reflection of nature, caught by ear or eye, and transfixed by the mirror of the human psyche. 'In a magnificent, but seldom read essay, Susanne Langer (1967) has suggested that all art is rooted in our response to natural form' (Stillman 1972). Others, from Goethe to our times, have agreed.

But today it is not enough to pay homage to human love for nature, nor to art's dependence on nature, but rather high time to press for experimental proof and scientific understanding of the 'why' and the 'how', and thus furnish the factual scientific-mechanistic bases and arguments for the deliberate preservation of the natural environment.

Crowded rats, growing brains, and denatured environments

Where are some of the frontiers in physiological research to redirect man's view of himself? What does this have to do with the biology teacher?

Human needs for nature surely cannot be disputed. Because we evolved in it, this, the biosphere, is the best of all possible worlds for us; and if we lose it, or even lose pieces of it, no one can predict the consequences. These cannot, in the long run, be good. Just to look at our urban problems is to glimpse into a hellish future darkly!

It is common knowledge that denatured human environments produce denatured (i.e. unnatural) humans. This is demonstrated over and over again in our crowded inner cities, our human zoos (see Morris, 1969), and especially well in the ultimate of denatured horrors invented by man—our sterile, crowded prisons. It is now equally well-established scientific knowledge that an unnatural rat environment produces unnatural (i.e. sick) rats (cf. Ward, 1972). In a series of classical experiments, John Calhoun of the National Institute of Mental Health showed how increased crowding in rat populations elicits maltreatment or abandonment of their young, as well as an increase in homosexuality and irrational violence, phenomena which are replicated by crowded *Homo sapiens* every day in the treeless slums of Chicago or under the smoggy skies of Los Angeles, and in a lot of nicer places, too.

The effects of crowding *per se* in man, irrespective of the nature of the environment, have not been as well understood or objectively studied. Recent tests (Ehrlich and Freedman 1971) suggest that under crowded conditions 'men become more competitive, somewhat more severe, and like each other less, whereas women become more cooperative and lenient and like each other more', an ethologically not unexpected finding. However, another study (Freedman, *et al.*, 1971) showed that people-density *per se* had no significant effect on performance of a variety of specific tasks carried out during four-hour periods on three consecutive days.

It may thus well be, as many technological optimists defending larger populations have hopefully repeated, that crowding alone, by and in itself, isolated from any concomitant side effects, exerts no particularly bad sociological influence on most people. (I personally doubt this, but for the sake of argument we may accept it here.) Why, then, are there so many of the most terrible modern human problems endemic to the crowded urban centres? If the cause is not crowding directly, perhaps it *is* crowding indirectly. Could it be, perhaps, that intense crowding, inevitably resulting in the total destruction of all complex biological ecosystems (allowing survival of a few species only, such as bacteria, fungi, certain weeds, insects, sparrows and rats) indirectly deprives man of a whole battery of inherited needs which can be satisfied easily and well only by nature? To put it in another way, since crowded places tend to be denatured, would this indeed not seem to, at least indirectly, support the ideas presented here: namely, that *humans cannot live very well without the biological diversity of nature*, and especially without the beauty and pattern of plants—to view, to explore, to be stimulated by. If it is nature and its diversity that has to be made available for man, the implications of this concept to biology teaching, to landscape architecture, urban planning, medicine, and to certain businesses, especially flori-culture and tree nurseries, are vast. A fundamental and synthesising, consciousness-raising re-evaluation of much of our literature is clearly in order. (cf. McHarg's 1969 maps of human pathologies concentrated in the heart of Philadelphia; see also Tinbergen and Tinbergen, 1972, a most remarkable ethological appraisal of autistic children, where, while the human interaction component is stressed, the natural needs for health are, if only barely, hinted at. Yet, one cannot help but wonder whether such overly timid children might not respond more readily to therapy in the uncrowded, unhurried, non-intimidating and more natural — for children especially — rural settings freely provided by village, farm or camp.

That 'poor', 'unstimulating', 'loveless' environments 'deprive' children, even without crowding, is likewise common knowledge, as do environments without the stimulation of some adversity, some rigour. But why should this be? How does this work? If we could demonstrate what is actually happening anatomically or physiologically to these understimulated children, might we not then be able objectively and realistically to plan a 'better' world for them?

Until recently, nothing was known about the *mechanisms* of deprivation, though much, of course, about the negative *results*. At long last, however, we are beginning to understand rather precisely what a 'poor' environment does, or rather does not do, for the mammalian organisms inhabiting it. In an article in the February 1972 issue of *Scientific American* entitled 'Brain Changes in Response to Experience', Mark Rosenzweig and his associates at the University of California in Berkeley summarised a series of experiments which demonstrated that baby rats raised in large cages with a diversity of stimulating objects turned out to become much more adept at solving problems than their siblings kept in smaller, duller cages.

A 'better' environment produced 'better' rats—of course! That in itself any biology teacher, in fact, almost any man in the street could have predicted. But Rosenzweig and his associates were able for the first time to show clearly in morphological, anatomical and physiological terms why these animals grow up to be more intelligent. By measuring, sectioning, and comparing their brains, they demonstrated a direct relationship between environmental diversity and the size and probable complexity of the brain (cf. Henderson 1970). In a stimulating environment, the cerebral cortex of the brains of baby rats grew larger by about 5 per cent to 10 per cent compared to those kept in a dull environment, a truly significant increase! Or to put it in another way, the ordinary and dull rat environment produced brains with cerebral cortices which were below normal by about 5 per cent. In effect, only in a rich or stimulating environment can the brains of baby rats reach their normal potential. Not only was brain size, brain weight, cortex thickness, etc. increased, but the individual nerve cells were shown to have grown more complex as a result of stimulation by diversity: with dendrites more branched and spines more abundant. Briefly, in the stimulating environment the nerve cells of the rat babies would 'fire' more often. This in turn somehow would cause them to grow more, to become more complex. Analogous to the enlarged biceps of the lumberjack, the brain 'adapts' by enlarging; or, again, to restate this in perhaps a more accurate way, the dull environment prevents the development of the highest or most optimal neural potentialities which, in a more natural environment, would ordinarily have been reached as a matter of course. This restatement makes it clear that, for baby rats, at least, diversity is not a luxury but biologically the pre-condition for normal development. Only a stimulated brain can become a fully developed and complicated (i.e. normal) brain! Demonstrated experimentally only for rats, it is probably an adaptation of all mammals.

If true for rat babies, is such stimulatory enrichment of brain potential (or suppression of potential) also true for human babies? That question, of course, cannot be answered with anatomical evidence, for such experiments with human babies cannot be done since mothers and fathers would object to having their offspring frozen, killed, and microtomed, even for a good cause! By analogy, however, (and perhaps by homology, too) what these tremendously significant studies clearly say is this: give a growing young mammal, from the earliest months on, a stimulating and healthy, more natural environment, and it will grow up to fulfil its innate potentialities, and develop a more complex, better and probably more harmoniously functioning and larger brain, and, perhaps, a healthier body and psyche as well. The animal will, in fact, become more intelligent! Applied to humans, it means brighter children, happier adults.

The cultural (as well as neurological) gap between humanity and other mammals, including rats, is of course immense. Rats, for example, are more nocturnal, with a more developed sense of smell and touch, humans more diurnal, with a much more developed sense of vision, natural pattern and sound. Because of our evolutionary histories, our needs differ in the *kinds* of stimulation we might need. Compared to all other animals, Man does stand alone, unique in a hundred ways—the walking, talking, writing, reading, car-driving, pair-bonding, fire-making, toilet-trained, social naked ape. Yet, even if we are very human, we still share all of our basic physical and psychological attributes with the other primates, and nursing and caring for the young with all other mammals. Thus, as remarkable as our species is, we must not continue to be blinded by our own self-image into believing we are at heart not animals with deep and innate needs and cravings which command so many of our actions. Therefore, the results of animal experiments *are* meaningful to teachers, to planners and to politicians. Indeed, our *basic* human patterns, both physiologically and behaviourally, and those of the primates hardly differ at all. One has only to read Jane Van Lawick-Goodall's (1971) account of her life with a tribe of chimpanzees in Africa to realise that both happiness and misery have universal roots in all the anthropoid apes, including man. *Especially for the young in their early ontogeny, happiness means satisfying basic innate natural needs*—and to babies the cultural satisfactions, which so many intellectualised adults insist on, are just so much extra whipped cream.

The evolution of intelligence as a function of diversity

In 'A Biologic View of Human History', a review of Seidenberg's (1950) memorable *Posthistoric Man*, the geneticist Bentley Glass (1951) says:

> The intelligence of man is an evolutionary product of natural selection for adaptability to great variation of surroundings, to tremendous vicissitudes of experience, as Bergson concluded forty years ago. It should follow from the nature of natural selection that the prolongation of the dominance of intelligence in man must depend upon the maintenance of the variety and richness, the vast complexity

of that world to which he must adjust his ways.

In the evolution of the human species, the vast diversity both
in immediate surroundings and in personal experience produced increasingly higher levels of intelligence through
natural selection, which was applied to cope with the
multiplicity of choices which such a diversity not only offers,
but imposes. Likewise, the potential for an increase in brain
complexity in response to diversity (especially during the
early life stages of a mammal, as shown by the above cited
experiments of Rosenzweig) may be viewed as a genetic
adaptation with survival value for those individuals which
find themselves, for whatever reason (and these in turn
probably largely based on genetic factors also), in a diverse
environment (i.e. more productive habitat) and which will
thus produce more offspring, and then brighter ones, than
those which settled in a dull, less productive environment.

Today, more than ever, it is evident that mankind needs
diversity and its stimulation not only to elicit the
anatomical-neurological responses of the individual growing
brain in childhood, but, on the global and phyletic scale, to
continue the high intelligence of the human species. As Glass
points out in the above review, just as environmental
diversity elicited the evolution of high intelligence in *Homo
sapiens*, so the general lack of diversity will most likely result
in a degeneration of this intelligence. The eyes of cave fish,
the legs of snakes and the useless muscles in the human ear
are cited as telling examples. It is a truism that, to maintain
any biological attribute, we need the continuation of the
environmental selective factors which produced it. We need
the stimulation of diversity to maintain high human intelligence, on both the personal and on the phyletic (species)
level, and, continues Glass, 'given an environment of diversity
and complexity, the evolution of intelligence should proceed'.

Yet what kinds of diversities? I realise that there are all
sorts, and civilisation produces its own unique brands. But
especially compelling and significant for man, I feel, must be
the incredibly rich diversity of the 400,000 species of plants
and four million or more species of animals which combine
and recombine endlessly into a vast array of perpetually and

astronomically variable biotic communities. From all indications, this visual natural diversity is especially meaningful to man, and here, at the earliest ontogenetic stages, almost compulsively searched for by children. It is well to stand in awe before this variability, which is of a magnitude quite irreproducible by technology.

Will mankind have enough sense to preserve the complexity and diversity of nature which may be necessary to induce, and to continue or preserve, high human intelligence? The current rate of heedless homogenation of the world's once diverse environments is not encouraging. The transformation of *natural* diversity into cornfields, cow pastures and concrete cities on a worldwide scale appears almost inevitable. Yet, concurrently, the marvellously interesting *cultural* diversity of the world is being obliterated as well. Eventually, human intelligence might well degenerate into stereotyped responses to the few stimuli allowed to survive.

Perhaps this evolutionary repression is already beginning in our larger cities where current unnatural selection pressures must be not only totally different from what they (until recently) were in more natural and benign environments, but are probably favouring human genotypes immune to ugliness, noise and pollution, and insensitive towards the life-sustaining systems of nature, since there, in a Chicago or a New York, these have all but ceased to exist. That is not to say that cities do not have compensating good qualities, especially for the socialising adolescent. But pity the young child who needs to see his first frog. Or, indeed, any sensitive or timid human in need of a slower pace, a more tranquil, natural way of living.

Continued 'growth' and crowding of course will inevitably increase these pressures. Growth both of population and of technology implies bigger and grimmer megalopolises and, if continued, condemns future generations (especially of the inner-city poor and black) to become permanently locked into permanently denatured and irrevocably polluted environments. The promise of an unlimited energy supply, especially, is the grimmest prospect of all; for it would postpone almost immediately any population curbs, increase tremendously industrial development and the 'bloating' of cities into megalopolises, and ensure the utter destruction of nature, and with it, ultimately, not only the source of our food and fibre but the sensitive and joyful nature of man.

If we value human intelligence, its superb sensitivities and deep emotions towards nature, and the happiness which flows from it, we had better value the natural diversity which still happens to exist by default or accident, and deliberately preserve it. Thus, by whatever route we approach the modern human problem, the answer is always the same: preserve nature, for it is sacred to humanity and its evolution.

Role of biology teachers

Why do people love flowers and shrubs more than rusty tin cans? Why do people insist on planting flower boxes and tilling vegetable and rose gardens even in the poorest sections of a town, often against impossible odds, and hire landscape architects at great expense in rich suburbs? Lead poisoning aside, why are there so many school problems with the deprived children of the poor, especially the blacks, in big city ghettos? And, finally, to beg a question, why must we now, at this late date, and at the last chance man will ever have, preserve the biotic richness of the lovely and good earth? Only through modern research may we arrive at definitive answers to these difficult and politically loaded questions.

How must biology teachers relate to the fantastic new discoveries of science? How could biology teachers rationally apply this increased understanding of the nature of man to their everyday activities?

Biology teachers could do much to become a major force for ecological good sense. They, more than other teachers, should have the knowledge of scientific principles on which this good sense must be based. They are capable of comprehending and accepting the fact that the animal origins of humanity are at its very base. They can appreciate the need for modern ethological and physiological research and be critical of the vagueness of many current precepts. They are aware that man is a slowly evolved and still slowly evolving primate species, and that any consideration of man in relation to his environment should be placed within the framework of his evolutionary nature. Biology teachers, especially, should be aware of the need for multidisciplinary studies of man, studies which must consider his original environment antedating the industrial and medical revolution, including the ethological predispositions of his tribal pre-nuclear family (cf. Greenbie 1972), the slowness of human evolution, the nature of human natural selection, as well as the great subtleties of chemical, physical and visual pollution effects and their relationships to sensory and psychological deprivation (Randolph 1962; Cassidy 1967; Iltis 1973a).

Many biology teachers, however, will need to undergo fundamental changes of attitudes; for the biology teacher as an intellectual revolutionary is not the image he generally projects. To encourage such rethinking and redoing, here are a few suggestions for action:

1 Above all, read critically the writings of the many biologists, environmentalists and social thinkers, such as Paul and Anne Ehrlich's *Population, Resources and Environment,* Barry Commoner's *The Closing Circle,* Meadows, *et al., The Limits to Growth,* Garret Hardin's *Population, Evolution and Birth Control,* Shepard and McKinley's *The Subversive Science* and *Environ/Mental,* or even Rachel Carson's *Silent Spring.* To get some idea of the limitations to man's technological use of this Earth, the report by the US National Academy of Science—National Research Council's Committee on *Resources and Man* should convince even the most optimistic whither our civilisation is drifting. The journal *Environment* in America, *The Ecologist* in Great Britain, as well as those of other environmental organisations belong in every biology teacher's home. Three exceedingly useful bibliographies also need to be mentioned, *On Rediscovering the Biosphere* by Dan McKinley (1970), *Science for Society,* edited by John Moore (1971) and *The Environmental Crisis: A Bibliography* by Carol Sherr (1973).

2 Read the writings of the new school of evolutionary perceptive, ethologically oriented anthropologists: the many sensitive books of Ashley Montague, Desmond Morris's *The Human Zoo* and *Intimate Behaviour,* Robert Ardrey's *African Genesis* and *The Territorial Imperative,* Anthony Storr's *Human Aggression,* Eibl-Eiblesfield's *Ethology,* Tiger and Fox's *The Imperial Animal,* Edward T. Hall's *The Hidden*

Dimension. Paul Shepard's masterful *The Tender Carnivore
and the Sacred Game*, works by Sommer, Lorenz, Tinbergen,
Elton, and even René Spitz's *The First Year of Life*. Not that
these authors will agree with each other, far from it. But you
may get a feeling for the immense force of the primitive
genetic components which interact with culture, and
especially for the sensitive innate needs of the human animal
during childhood. For the lower level teacher, the short but
excellent *Basic Ecology* text of the Buchsbaums (1957),
written clairvoyantly before the environmental crisis, is still
to be highly recommended. Modern textbooks dealing with
the general subject of human ecology and evolution will be
most rewarding (e.g. Odum 1971; Boughey 1972; Ehrlich *et
al.*, 1973) but, in a sense, *every* book listed so far deals with
these topics.

It is crucial to realise that most of the authors here listed
ignore or underplay human genetic needs for nature, or at
best tend to confuse it with cultural sophistications of the
elite. Very few have a holistic view of genetic programming,
cultural training, and environmental input as an indivisible
unity (cf. Bateson, *Steps to an Ecology of Mind*). And very
few indeed are truly concerned with, or even aware of, the
most irreplaceable and irreversible, tragic, and, yet, ultimately
most stupid effect of the exponential growth of human
activity, population, and affluence: the current extinction of
plant and animal species and ecosystems on a gigantic global
scale, especially now in the tropical rain-forest regions, and
soon in all the deserts and oceans. David Ehrenfeld's superb
book, *Preserving Life on Earth*, Dasmann's *A Different Kind
of Country*, Dorst's *Before Nature Dies*, the beautiful *Wildlife
in Danger* by Fisher, Simon and Vincent, the many publications of the International Union for Conservation of Nature
and Natural Resources (IUCN), including its excellent
bulletin, and some of my own papers speak to that point. But
it is a little hard to comprehend why the otherwise valuable
semi-official guide book for the United Nations Stockholm
Conference on the Human Environment, *Only One Earth*, by
Barbara Ward and René Dubos, hardly mentions extinction
of species or of diversity, an almost inconceivable yet not
really surprising oversight, considering that most authors and
readers today are city-born and city-bred. Even Konrad

Lorenz' latest book, on the eight deadly sins of modern man, singles out growth, overpopulation and pollution, but the ruthless eradication of ecosystems and the resulting extinction of plants and animal species is not one of them. Why are we, even renowned biologists, so reluctant to mourn the loss of species, this incredible, incalculable and, from most any perspective, insane calamity?

3   The need for nature is yet to be objectively documented. There are a multitude of *subjective* indications, from religion, art, and music to gardening and psychiatry, that suggest man to be indivisible from the living natural environments around him. On the other hand, highly *objective* indications, such as the physiological-anatomical work cited above, are rare indeed. Clearly, more information, more research is desperately needed on man's perception of the whole living, blooming, fruiting plant, and animal, both in the landscape and in his kitchen, before it is too late. While *Plants/People/ and Environmental Quality* (Robinette 1972) has little to say on man's genetic needs for plants or nature, it does deal thoroughly with the architectural, engineering and climatological uses of plants, and includes a very extensive bibliography on that aspect. Paul Shepard's *Man in the Landscape* is one of the few books that delves deeply into our fundamental physical and psychological adaptations to nature. There is now a vast array of books, symposia and anthologies dealing with the environmental crisis, many of which will give isolated perceptive insights into this question (e.g. the preface to Fisher *et al.*).

Lewis Mumford's *The Myth of the Machine*, and others of his books enable one to see technology not only as a servant but as a master of humanity. It might not be amiss to point out that there is now a compendium of failures of technology. On the large scale, and in the long run, it just doesn't work! *The Careless Technology*, edited by Farvar and Milton, 'is a book full of failures, and failures of the most vivid sort, in fields that are at the heart of what we think we know most about: building dams, fighting disease, making food available, etc. It is the Green Revolution a dozen times over', (Dan McKinley, personal comment). Paul Goodman's essay *Can Technology be Humane?* insists on prudence and on the need, not only for a change in the values and organisation of science, but 'in the kinds of men who make scientific decisions'. Roderick Seidenberg's difficult *Posthistoric Man* is hardly optimistic about the outcome of human evolution, especially if we do not consciously take alternatives to a technology which makes human activities, and even human needs, superfluous. René Dubos's *A God Within, Man Adapting* and *So Human an Animal* contain valuable bio- logical observation of man's intimate responses to nature, yet here, and especially in many recent papers Dubos (1973) has turned into a kind of environmental 'Uncle Tom', taking potshots at those concerned with the preservation of species and ecosystems. In some of my papers, I have cited Maxwell Weissman (1965), a psychiatrist, who found much improve- ment in mentally ill patients being taken to a rural natural setting. Others have made similar observations without follow-up research. Our lack of hard knowledge on nature as a healing environment is but a reflection of our almost deliberate and cynical neglect of studying the man-nature relationship, despite the fullness of our mental hospitals and the many other pathological manifestations of a denatured and sick world.

4   Each biology teacher is on his own to comprehend the evolutionary nature of mankind and synthesise his own philosophy. Hopefully, this will be based on prudence, restraint, respect for innate needs, and a sincere desire to reduce the impact of humanity on the earth. Perhaps nothing is more important today than this. We as individual teachers have to develop our *own* biological sophistication, or we are lost. What is one to think of pontifical projections of an *optimum*(!) world population of 50 billion (the figure given by some Russian planners, by the Australian Colin Clarke and the Greek planner Doxiadis), 36 billion (the FAO), or even a mere nine billion (by none other, of all people, than the very man who showed what crowding did to rats; cf. Calhoun, 1968), when today, the world in 1973 is heading into unprecedented famine with not quite four billion, and when in fact a total world population of one billion would be more than an ample breeding population and be capable of doing anything the human species would ever want to do? How is

one to deal with Krieger's *What's Wrong with Plastic Trees?* or any of the legion of Utopian environmental optimists which lately have found such favourable editorial reception in *Science* and *Nature*, except by ridicule? (Iltis 1973b). It is obviously important to examine critically the 'glad tidings' of these optimists who will not accept the premise that *a finite world has finite resources*, including plants and animals; who readily promise, even today, the pie-in-the-sky of unlimited resources, unlimited food, unlimited oceans, unlimited atomic energy, unlimited population, and unlimited wealth, but who neglect to mention the unlimited misery humanity is heading into as a consequence, or the unlimited extinction of plant and animal species. Newspapers and magazines, likewise, abound with articles making self-fulfilling prophesies: the 'need' for *more* growth, *more* electric power, *more* food, *more* roads, and *more* of nearly everything. It is frustrating and galling to observe that in this 'scientific' age, everyone seems to be culturally conditioned to demand exacting proof of human 'needs' for nature, while no one ever questions the 'need' for more cars, more cows, or more concrete, needs that in no sense are biological.

Remember, nobody ever can get something for nothing. 'There is no such thing as a free lunch' to quote Barry Commoner's '4th Law' of ecology. Atomic power without polluting radiation; Green Revolutions without depletion of genetic diversity, without pesticides, or without disastrous plant diseases; population expansion without extinction of plants and animal species; these are the pipe dreams out of which the ultimate human tragedy is now being deliberately constructed. Thus, it *is* true, as a brilliant if sick joke will have it: 'If you want a higher standard of living, you will have to settle for a lower quality of life!'

Finally, if an all-encompassing eco-evolution is indeed on the horizon (and it had better be if we are to survive with any sanity), it must be based on a thoroughly ecologically-sound restructuring of society to ensure the survival of biotic diversity. The human species, once free and frivolous in its habits of consumption, must be brought into a steady-state harmony with the natural environment. The limitation of population size and consequent restructuring of economics and priorities is the fundamental prerequisite for a meaningful long-term resolution of the environmental crisis. The ultimate aim for our species must be to find the basic optimum human environment: a compromise between our cultural accomplishments and our biological needs—a compromise between many of the advantages of technological urban civilisation, which have freed man's hands from labour and man's brains for creativity and which we are loath to give up; and the innate and often subtle human needs for the evolutionary theatre of nature, the wild environment which produced our bodies and brains over millions of years through natural selection; needs which we cannot ignore except at gravest peril to our health, sanity, and happiness.

The world is changing at a terrifying rate, but human needs for nature will remain essentially the same, for millennia and decimillenia to come. Even without knowing the precise physiological or neurological details on which these needs are based, we must respect them as the forces which have moulded our species both physically and psychologically. It shall be the true role of biology to make our life livable and human by insisting that our inherited needs for natural pattern, natural form, and natural diversity are not abandoned wherever we may be in favour of a capricious technological hell. In the preservation of nature lies the salvation not only of biology teaching, but of humanity as well.

References

R. Ardrey (1961) *African Genesis*, Collins.
R. Ardrey (1967) *The Territorial Imperative*, Collins.
R. Carson (1965) *Silent Spring*, Penguin.
B. Commoner (1972) *The Closing Circle*, Cape.
R F. Dasmann (1968) *A Different Kind of Country*, Macmillan, New York.
J. Dorst (1970) *Before Nature Dies*, Collins.
R. Dubos (1965) *Man Adapting*, Yale University Press.
R. Dubos (1970) *So Human an Animal*, Hart-Davis.

R. Dubos (1972) *A God Within*, Scribner's.

R. Dubos (1973) 'Humanizing Earth', *Science*, 179.

D. Ehrenfeld (1972) *Preserving Life on Earth*, Oxford University Press, New York.

P. R. and A. H. Ehrlich. (1972) *Population, Resources and Environment: Issues in Human Ecology*, 2nd edition, W. H. Freeman.

M. T. Farvar and J. P. Milton (eds.) (1973) *The Careless Technology*, Tom Stacey.

J. Fisher, N. Simon and J. Vincent (1969) *Wildlife in Danger*, Viking Press.

P. Goodman (1969) 'Can Technology be Humane?', *New York Review of Books*, 20 November, 1969, reprinted in *The Social Responsibility of the Scientist*, edited by Martin Brown, 1971, Free Press.

T. Hall (1966) *The Hidden Dimension*, Doubleday.

G. J. Hardin (ed.) (1964) *Population, Evolution and Birth Control: a Collage of Controversial Ideas*, W. H. Freeman.

M. H. Krieger (1973) 'What's wrong with Plastic Trees?', *Science*, 179 (4072). An edited version 'Up the Plastic Tree' reprinted in *Landscape Architecture*, 63 (4) (1973).

S. Langer (1967) *Mind: An Essay on Human Feeling*, vol 1, Johns Hopkins University Press.

A. Leopold (1949) *A Sand County Almanac*, Oxford University Press, New York.

K. Lorenz and P. Leyhausen (1973) *Motivation of Human and Animal Behaviour*, Van Nostrand Reinhold.

D. McKinley (1970) 'On Rediscovering the Biosphere', *The Explorer*, 12 (2).

D. H. and D. L. Meadows, J. Randers and W. W. Behrens (1972) *The Limits to Growth*, Universe Books.

J. A. Moore (ed.) (1971) *Science for Society: a Bibliography*, revised edition, American Association for the Advancement of Science.

D. Morris (1969) *The Human Zoo*, Cape.

D. Morris (1971) *Intimate Behaviour*, Cape.

L. Mumford, (1967) *The Myth of the Machine*, vol.1, *Technics and Human Development*, Secker & Warburg.

L. Mumford (1971) *The Myth of the Machine*, vol.2, *The Pentagon of Power*, Secker & Warburg.

G. O. Robinette (1972) *Plants/People/and Environmental Quality, a study of Plants and their Environmental Functions*, U.S. Government Printing Office.

M. R. Rozenzweig, E. L. Bennett and M. C. Diamond (1972) 'Brain Changes in Response to Experience', *Scientific American* 226 (2).

R. Seidenberg (1950) *Posthistoric Man: an Inquiry*, University of North Carolina Press, reprinted in 1957 by Beacon Press.

P. Shepard (1967) *Man in the Landscape: a Historic View of the Esthetics of Nature*, Knopf.

P. Shepard (1973) *The Tender Carnivore and the Sacred Game*, Scribner's.

C. Sherr (1973) *The Environmental Crisis: a Bibliography* in Lange, C.

T. and Klinge, P. E., see Further Reading.

R. Spitz (1965) *The First Year of Life*, International University Press.

C. W. Stillman (1972) 'Reflections on Environmental Education', *Teachers College Record*, 74 (2).

A. Storr (1969) *Human Aggression*, Allen Lane.

J. Van Lawick-Goodall (1971) *In the Shadow of Man*, Houghton Mifflin.

B. Ward and R. Dubos (1972) *Only One Earth: the Care and Maintenance of a Small Planet*, Penguin.

Further Reading

G. Bateson (1973) *Steps to an Ecology of Mind*, Ballantine.

A. S. Boughey (1971) *Man and Environment: an Introduction to Human Ecology and Evolution*, Macmillan, New York.

M. and R. Buchsbaum (1957) *Basic Ecology*, Boxwood Press.

J. B. Calhoun (1968) 'Space and the Strategy of Life', paper presented at The American Association for the Advancement of Science, Dallas.

H. G. Cassidy (1967) 'On Incipient Environmental Collapse', *BioScience*, 17.

R. M. Chute (1971) *Environmental Insight: Readings and Comment on Human and Nonhuman Nature*, Harper and Row.

C. A. Doxiadies (1966) *Urban Renewal and the Future of The American City*, Public Administration Service, Chicago.

P. R. and A. H. Ehrlich and J. P. Holdren (1973) *Human Ecology: Problems and Solutions*, W. H. Freeman.

P. Ehrlich and J. Freedman (1971) 'Population, Crowding and Human Behavior', *New Scientist and Science Journal*, 1 April, 1971.

I. Eibl-Eibelsfeldt (1970) *Ethology*, Holt, Rinehart and Winston.

Environment, vols. 1–15 (1973), P.O. Box 755, Bridgeton, Missouri 63044.

J. W. Forrester (1971) *World Dynamics*, Wright-Allen Press.

J. Freedman, P. Ehrlich and S. Klevansky (1971) 'The Effect of Crowding on Human Task Performance', *Journal of Applied Social Psychology*, 1 (1).

B. Glass (1951) 'A Biologic View of Human History', *Scientific American*, December (review of Seidenberg's *Posthistoric Man*)

B. Greenbie (1972) *Homo Sapiens Habitat: Implications of Ethology for the Design and Planning of Human Habitations*. PhD thesis, University of Wisconsin.

N. D. Henderson (1970) 'Brain Weight Increases Resulting from Environmental Enrichment: A Directional Dominance in Mice', *Science* 169.

J. P. Holdren and P. Ehrlich (eds.) (1971) *Global Ecology: Readings Toward a National Strategy for Man*, Harcourt Brace.

H. H. Iltis (1959) 'We Need Many More Scientific Areas', *Wisconsin Conservation Bulletin*, 24(9).

H. H. Iltis (1966) 'The Meaning of Human Evolution to Conservation',

Wisconsin Academy Review 13(2).

H. H. Iltis (1967) 'To the Taxonomist and Zoologist: Whose Fight is the Preservation of Nature?' *BioScience* 17(12).

H. H. Iltis (1968) 'The Optimum Human Environment and its Relation to Modern Agricultural Preoccupations', *The Biologist* 50(3–4).

H. H. Iltis (1972) 'Conservation, Contraception and Catholicism, A 20th Century Trinity', *The Biologist* 54(1).

H. H. Iltis (1972) 'The Biology Teacher and Man's Mad and Final War on Nature', *The American Biology Teacher* 34(3).

H. H. Iltis (1972) 'The Extinction of Species and the Destruction of Ecosystems', *The American Biology Teacher* 34(4), April.

H. H. Iltis (1973a) 'Pollution and Adaptation: What Hope for Man' in C. T. Lange and P. E. Klinge (eds.) *Pollution*, 1–6, (see below).

H. H. Iltis (1973b) 'Down the Technological Fix', (original title: 'Can One Love a Plastic Tree?' an answer to Krieger 1973), *Landscape Architecture* 63(4).

H. H. Iltis, O. L. Loucks and P. Andrews (1970) 'Criteria for an Optimum Human Environment', *Bulletin of the Atomic Scientists* 25(1).

C. T. Lange and P. E. Klinge (1973) *Pollution*, A special publication of the National Association of Biology Teachers, 1420 N Street, N. W., Washington, D.C., 20005.

National Academy of Science-National Research Council (1969) *Resources and Man*, W. H. Freeman.

H. T. Odum (1971) *Environment, Power, and Society*, Wiley-Interscience.

E. P. Odum (1971) *Fundamentals of Ecology*, 3rd Edition, W. B. Saunders.

T. Randolph (1962) *Human Ecology and Sensitivity to the Chemical Environment*, Charles C. Thomas.

P. Shepard and D. McKinley (eds.) (1969) *The Subversive Science: Essays towards an Ecology of Man*, Houghton-Mifflin.

P. Shepard and D. McKinley (eds.) (1971) *Environ/Mental: Essays on the Planet as a Home*, Houghton-Mifflin.

L. Tiger and R. Fox (1972) *The Imperial Animal*, Secker and Warburg.

The Ecologist : 'Man and the Environment', 'The Quality of Life', 'Pollution', 'Conservation', vol.1 (1971) from 73 Kew Green, Richmond, Surrey.

E. A and N. Tinbergen (1972) *Early Childhood Autism—An Ethological Approach*, Fortschritte der Verhaltensforschung, Beihefte zur Zeitschrift für Tier psychologie 10 (Advances in Ethology. Supplement to *Journal of Comparative Ethology*, 10, Verlag Paul Parey, Berlin und Hamburg.)

Agents of change?

CYRIL SELMES

Introduction—The credo of a biology teacher

It is probably true to say that nowhere else in the world does a teacher have so much freedom in the choice of what to teach as he does in Britain.   (Wiseman and Pidgeon, 1972)

Too much of the tradition of science teaching is of the nature of confirming foregone conclusions. It is a kind of anti-science, damaging to the lively mind, maybe, but deadly to the not so clever.   (Newsom Report, 1963)

It is a paradox that those of us who are educators and by implication leaders of the population, are literally a limiting factor in the educational process. It is time for a soul-searching inquiry in which we reassess our objectives and ask whether our programme is realistic in enabling students to achieve these objectives.'
                    (Postlethwait, Novak and Murray, 1969)

...the 'nuts and bolts' of a new science curriculum are considered the most important 'to-be-learned' aspects of the ... regime. To wit: familiarity with the *text*, the *laboratory manual*, the *teacher's guide*, and requisite *equipment* and *supplies*. After all—this is where 'the action is', or is it?'
                    (Cohen, 1972)

These four quotations are intended to focus attention on the main arguments of this chapter and to indicate, in part at least, my *credo* as a teacher. When everyone has had their say—heads of departments, headmasters, examining boards, parents, inspectors, curriculum developers, curriculum evaluators, teachers' unions, professional associations, and many others—it is the individual teacher who decides *what* and *how* he (or she) will teach. Nothing the teacher may say—'I don't have enough time for my work', 'the laboratory facilities are awful', 'the GCE syllabus restricts my work'—can absolve him from this responsibility. This is not to deny that time, money, facilities, equipment and examinations are important factors which influence the work of a teacher, but to suggest that preoccupation with these matters may be used as an excuse, or even a justification, for what occurs in the classroom or laboratory. Certainly it could be argued that the pace and fervour of new projects, which have been characteristic of curriculum development during the last decade, is the result of cumulative dissatisfactions with what is taught and how it is taught. And these projects, in their turn, have made clearer some of the probable causes of earlier dissatisfactions and some of the results of piecemeal curriculum reforms. For example, we are beginning to realise the need for systematic thought about the principles of curriculum development (Wiseman and Pidgeon, 1972), the necessity for these ideas to become part of the working knowledge and skills of practising teachers (Whitfield, 1971), and the danger of being influenced by emotive language and the attraction of new labels on old bottles (Wiseman and Pidgeon, 1972). The one I wish to emphasise, however, is the tendency for teachers to be almost exclusively concerned with the content of their courses (Taylor, 1970), and their relative lack of concern for either the aims or the evaluation of their work. If this is true of the majority of teachers 'on the shop floor' and if this 'nuts and bolts' approach is also typical of both initial and in-service training of teachers, then there may be a cause-effect relationship from which it is difficult to escape. In other words, concern for content during degree work may be reinforced by concern for content during training as a teacher, and this will be further emphasised by both the secondary school environment and in-service training.

All these comments are applicable to the teacher of biology. If he is to make a contribution to the curriculum of a school—all the experiences for learning which are planned and organised by the school (Kerr, 1968)—he must be able to present a reasoned case for the inclusion of biology in the curriculum and a detailed defence of the particular syllabus in use. He should also be able to give some justification of the teaching methods to be used, even if it is difficult to evaluate either the content or the methods.

The main purpose of this chapter is to discuss the case for

including biology in the curriculum and to suggest some ways in which ideas about the aims (or goals) of biology teaching may be clarified, a necessary prerequisite for the production of course content and eventual evaluation. Most of the suggestions are applicable to biology teaching in Further and Higher Education as well as secondary schools.

How can a biology teacher begin to build up a reasoned case for biology? He must concern himself with attempting to state:

1 the unifying themes (or principles, or concepts) of biology;
2 the processes by which knowledge of biology is obtained;
3 the values and attitudes which are essential to the progress of biology.

Principles of biology

One of the first problems in attempting to state the unifying themes of biology is one of terminology. As Novak (1965) has pointed out, there are various terms in the literature of science teaching which are used as if they were synonymous, for example, concept, conceptual scheme, theme, organisational thread, major generalisation, major concept, fundamental idea, and major principle. When an individual teacher is involved in this exercise he can choose his own definition; the problem is more acute when a group of teachers is involved, but this may be resolved by asking individuals to provide examples of what is thought to be a theme or principle.

The second problem is one of habit: many teachers (or prospective ones) react to this question by listing the topics (or subdivisions) of biology with which they are familiar, for example, photosynthesis, transpiration, respiration, osmosis, and so on. These are often the subdivisions or headings under which they were taught (or expected to learn) some of the subject matter of biology. This approach may have been adopted for excellent reasons but it may not contribute to a unified picture of biology.

Some examples may help to clarify the concept of theme or principle. One of the simplest statements of the principles of biology was written by Preston (1936):

1 Living things differ from non-living things in that they possess a definite cellular structure—the principle of *cellular organisation*.

2 There is a constant building-up and tearing-down within the body of every living organism—the principle of *metabolism*.

3 Living things grow from within through changing food into living protoplasm and new cell walls—the principle of *growth*.

4 Living things must adapt themselves to the conditions of their environment if they are to survive—the principle of *adaptation* or ecology.

5 The varying success of organisms in adapting themselves to various environmental conditions has given rise to topographical, geographical and climatic groupings—the principle of *distribution*.

6 All living things have the power to produce new organisms similar to themselves—the principle of *reproduction*.

7 Offspring inherit parental characteristics—the principle of *heredity*.

8 All forms of life on earth today have developed through the ages by gradual changes influenced by both heredity and environment, as the progeny of some simple primordial organisms—the principle of *evolution*.

9 Organisms past and present can be grouped according to one 'family tree' which shows what is believed to be their hereditary or evolutionary relationships—the principle of *classification*.

10 The interdependence and interrelations of living things give rise to a natural equilibrium, the temporary disturbance of which takes the form of epidemics—the principle of *balance in nature*.

11 All living things react to external stimuli—the principle of *behaviour*.

Another attempt was made by the National Society for the Study of Education (1932):

1 Energy cannot be created or destroyed, but merely transformed from one form to another.
  (i) Energy is supplied to organisms by oxidation processes.
  (ii) A continual food supply is required to provide energy.
  (iii) Because protoplasm contains protein, this must be supplied in the food sources.
  (iv) Food substances must not contain deleterious matter.
  (v) Oxidations must result in waste metabolites which have to be removed.
2 The ultimate source of energy of living things is sunlight.
3 Micro-organisms are the immediate causes of some diseases.
  (i) Transfer of micro-organisms can be prevented.
  (ii) Microbes exhibit differential survival capacities.
4 All organisms must be adapted to the environmental factors in order to survive in the struggle for existence.
  (i) There are complex interrelationships between organisms.
  (ii) Man must protect those plants and animals which have been domesticated to ensure their survival.
5 All life comes from previously existing life and reproduces its own kind.
  (i) The raw organism starts life as a fertilised egg.
  (ii) Unit characters are inherited and determined by genes.
  (iii) There is a mechanism for distribution of characters.
  (iv) Some characters depend upon more than one gene.
  (v) Proximity of genes leads to linkage.
  (vi) Crossing-over may break up link-associations.
6 Animals and plants are not distributed at random, but are found in definite zones and societies.
  (i) These zones are separated on the large scale by geographical barriers.
  (ii) Local societies depend upon micro-factors of climate and food.
7 Food, oxygen and certain optimal conditions of temperature, moisture and light are essential to the life of most

living things. (Undesirable organisms are often destroyed by controlling one of these factors.)

8 The cell is the structural and physiological unit in all organisms.

9 The more complex organisms have been derived by natural processes from simpler ones, these in turn from still simpler, and so on back to the first living form.

What is the relevance of these lists to the teaching of biology?

If these principles are important, any course of biology which ignores one or more of them will give an unbalanced picture of biology. If these principles are important, all children should be introduced to some of the information which has led to these generalisations and should be helped to understand them.

Although both lists show considerable agreement about the principles of biology, they were both produced some time ago, before the relative explosion of biological knowledge. A more up-to-date attempt to define the major concepts of biology has been made by Thompson and Pella (1972) who asked a national panel of biologists, science educators and secondary school biology teachers: 'What biology concepts are important to a student graduating from high school?' Full details cannot be given here as a final list of 114 concepts was produced, each of which was considered important by the panel. The twelve most important concepts were:

1 With minor exceptions, living things obtain their energy directly or indirectly from the sun through the process of photosynthesis.

2 All living things are interdependent with one another and with their environment.

3 The natural environment is being so altered through the effects of technology and the rapid expansion of wealth and population that man must make drastic changes in his behaviour or consider his extinction probable.

4 To meet future needs, the use of biological resources must be governed by the role these resources play in the ecosystem.

5 Living things tend to increase in numbers to the level

the environment will permit. Man's ability to modify his environment does not make him an exception. His success in by-passing some environmental barriers is now challenged by new barriers.

6 By and large living things are composed of a fundamental unit of structure and function known as the cell.

7 No one major ecological factor but a combination of them determines the environment; soil, water, air, energy, plants, and animals (including man) all contribute.

8 The earth's carrying capacity is finite for one and all species. Its space, materials, and available energy are limited mainly by surface area, the rate of cycling of water, gases, and materials, and the efficiency of photosynthesis.

9 The phenomenon of heredity in all living things thus far investigated is attributable to the replication and transmission of the genetic materials DNA and RNA.

10 Present living cells are seen to arise from pre-existing living cells.

11 In a biochemical reaction there is no loss or gain of energy, only a transformation of energy from one form to another.

12 Energy passes through a network of organisms beginning with green plants—the food producers—then to animals which eat plants, then to animals that eat those animals and so on. At each transfer energy is dissipated and less life can be sustained.

This list of concepts (or unifying themes) shows some overlap of content or subject-matter: this may be due to the composite way in which the list was obtained, by getting the collective opinions of a large number of people (387) working independently. In other words, this list was not produced by an individual, like Preston, trying to summarise the principles of biology, nor by a group, like a committee of the National Society for the Study of Education, trying to provide a consensus opinion about the principles of biology. In spite of this, the list still includes many of the principles stated by Preston and many of the concepts enunciated by the NSSE. For example, it includes a statement of the source of energy for living things (no.1 above); varying statements of the principle of balance in nature (nos.2, 3, 4, 7, 8); a principle of cellular organisation (no.6); a principle of heredity (no.9); a principle of reproduction (no.10); and two statements concerned with energy transfer, metabolism and growth (nos.11 and 12). The other difference between this list and the previous (older) lists is an emphasis on the environment, in general, and man's involvement with this environment in particular. This may reflect our present concern with problems of pollution and conservation and our growing awareness of the interaction between man and his environment.

How does this consideration of the principles of biology help the teacher to present a reasoned case for biology?

A teacher who engages in this kind of activity is reaching the stage where he can say:

these are some of the fundamental ideas which biologists have elucidated since the study of living organisms began; and:

it is understanding of these ideas which should result from the teaching and learning of biology.

The teacher is thus beginning to describe some of the characteristic features of biology which may justify its inclusion in the curriculum and to suggest that understanding of these principles should be transmitted to the next generation: in other words, *cultural* reasons for the teaching and learning of biology are implied.

Other cultural reasons may be advanced for the study of biology: unless some understanding of these concepts is encouraged *in* school, children have little chance of:

1 understanding future developments in biology, some of which may have important repercussions on their everyday lives;
2 developing an organised body of knowledge related to living organisms which can be used to interpret new experiences;
3 being able to communicate with each other—and with adults—about living organisms and biological concepts.

These statements lead to another justification for biology teaching—a *utilitarian* one. Many of these concepts (or the biological information encompassed by them) are important because they are related to practical considerations which affect both society and individuals; biological knowledge has led to control. For example, establishment of the principle of biogenesis (reproduction, or like begets like) by Pasteur and subsequent workers has led to understanding of the control of micro-organisms, both beneficial and harmful ones; establishment of the principle (or mechanism) of heredity has led to control of the breeding of more productive varieties of both plants and animals; and knowledge of the interrelations between living organisms has enabled Man to eliminate other species of animals competing for these increased food supplies; sometimes, it is now realised, with unexpected side-effects. Not only is this kind of knowledge useful to society in general but also to individuals. The concept of biogenesis is applicable to many everyday events: beneficial organisms (yeasts used in wine or beer making) can only be produced by previously existing organisms and thus it is necessary to introduce these organisms into the ferment either naturally (by crushing the grapes) or artificially (by buying the required variety); harmful organisms (disease-causing bacteria or viruses) can only arise from previously existing organisms and thus it is possible to control the transmission of these organisms from one person to another; organisms (bacteria and fungi) which grow on food can only arise from previously existing organisms and thus it is possible to prevent the access of these organisms to food. Other concepts may also be applied to everyday events in a similar way.

Thus many of the principles or concepts are important for both cultural and utilitarian reasons. Other reasons for biology teaching result from considering the processes by which biological knowledge is obtained.

Processes of inquiry in biology
Biology is a scientific discipline and therefore asks questions which are characteristic of a science. Eggleston (1971) suggests there are two kinds of questions apposite to biology: 'what' questions which involve observation and classification,

and 'why' questions which involve the formulation of hypothetical models in order to explain links between events. In order to answer these questions, biologists (like other scientists) undertake scientific investigations. These often involve one or more of the following processes:

1 making observations, classifying them and speculating about them;
2 producing hypotheses in order to relate or explain sets of observations and to establish causal relationships;
3 testing these hypotheses by further observations which are planned to occur under experimental conditions;
4 planning experiments in which the evidence is collected in a systematic and objective way under conditions which may be repeated and which produce similar results;
5 interpreting the data, either qualitative or quantitative, and estimating the experimental error;
6 reconsidering and modifying hypotheses in the light of experimental results, and making further predictions about future experiments.

Why is this kind of list important?

If these are considered to be important aspects of biology, any course which ignores them will give an unbalanced picture of biology. Thus teachers should provide opportunities for children to engage in these processes or activities which involve them.

Certainly this aspect of biology teaching has been emphasised since the 1930s, particularly in the writings of the Science Masters' Association (1938, 1947, 1950, 1953)—now the Association for Science Education—but it has become prominent only as a result of the Nuffield Science projects. It is difficult to judge, however, the effect of this increasing concern with biology as a process of investigation on either biology teachers or biology teaching. My personal impression is that many teachers still consider the subject matter of biology more important than the processes of investigation. This may still be due to the influence of external examination syllabuses and examination papers which continue to place a premium on memorised material (see chapter 16). It may also be due to teachers who have not participated in biological investigations since they left their university or college of education, or have not continued to consider the nature of scientific investigation. Consequently, when investigations are introduced, they are presented as a rather routine process where the emphasis is placed on the careful handling of apparatus and the logical nature of scientific thought. There is some evidence (Selmes *et al*, 1969) that it is precisely this emphasis on the cold, impersonal, rational process of investigation which discourages children from continuing the study of science. They feel there is little opportunity to express their own ideas or to use their own initiative. Contrast this with the enjoyment, exhilaration and excitement of scientific investigation as described by Snow (1958) in fiction and by Watson (1970) in real life. And many less well-known researchers would vouch for the 'excitement of the chase' and both the mental joy and anguish of research.

In other words, more than the processes listed above are involved in investigation. Beveridge (1961) discusses the part played by intuition, imagination, and chance or luck. Medawar (1964) questions the rationale of writing scientific papers which exclude personal speculation, and other essays in this book provide a variety of examples of the personal involvement and satisfaction which are derived from attempts to solve a scientific problem. Jevons (1961) provides a comprehensive discourse on science in many of its aspects—as a body of knowledge, an investigatory process and a social activity—and also suggests that the Baconian model of generalisation by logic from objective observation has prevented (and may still be preventing) the understanding of science by non-practitioners. In addition, some of the psychological difficulties involved in inquiry processes are becoming clearer. The selection and distortion of observations during perception throws doubt on the possibility of objective observation, without constant checking of both the observer and his use of instruments (Bartlett, 1951); the limited range of *schemata* (patterns of organised knowledge) with which we habitually interpret our surroundings suggest we may be both inefficient producers of alternative hypotheses and inefficient interpreters of data (Johnson-Abercrombie, 1960); and the prevalence of stereotypes of people and institutions, together with attitudes which are emotionally-based, may make it difficult for teachers (as well as children) to consider science in a different way (Hudson, 1970).

In spite of these difficulties, the teaching and learning of biology should involve not only the rational processes of scientific investigation, but also the non-rational ones. If these non-rational processes are essential features of investigation, children must, in some way, be introduced to the thrill and excitement, made aware of the part played by imagination and luck, and led to some understanding of the limitations imposed by their physical and mental make-up.

Three kinds of reasons have now been presented for the inclusion of biology in the curriculum: cultural, utilitarian and investigatory ones. If we are successful in putting these ideas into practice, what kind of person might result from these activities?

Values and attitudes of biology (or biologists)
Biologists must not only be familiar with some of the subject matter of biology but also with the process of inquiry. As a biologist engages in these processes—intellectual, rational and non-rational—certain characteristics, (or abilities, or attributes) develop; at least, it is usually assumed that a biologist will have characteristics different from other scientists as a result of their differences in education. Some of these characteristics or abilities might be:

1 honesty and integrity in the planning, performance and communication of his investigations;
2 open-mindedness in his approach to problems and in his production and testing of hypotheses, accompanied by an avoidance of dogmatism and a recognition of the need for verification of results;
3 the ability to work independently and yet, when necessary, to cooperate with other workers;
4 capability to evaluate the implications of his work for human society.

To what extent *do* these characteristics appear in biologists? From a sociological viewpoint, it might be argued that recognition as a biologist will not be given by the scientifio

community unless he provides evidence, usually by publication but sometimes by verbal communications, of the work
he has done and the results obtained (Cotgrove and Box,
1970). Dishonesty, lack of integrity, dogmatism and carelessness in verification would separately, or severally, hamper
such recognition. Yet many useful investigations have
resulted from biologists who have obstinately held to a
hypothesis in the face of opposing evidence (see Beveridge).
Moreover, like other scientists, biologists appear to prefer
working in isolation and to shun situations which involve
people (Roe, 1953). Biologists have also been perceived by
some young people (Hudson, 1970) as not particularly
intelligent, less valuable than other scientists and rather
unreliable. Biology teachers may also be included in the
stereotype of science teachers which emphasised the cold,
dull, dreary, uninteresting and sexless life they led.

Assuming, however, these characteristics *do* result—what
other values will a biologist possess? He will probably believe
that the study of biology is a 'good thing' in itself, perhaps
for a variety of reasons—for example, it may have been a
constant source of interest and satisfaction, or a starting
point for an all-absorbing hobby, or even the means to obtain
a steady income. He may also feel that biology is an
important and useful subject because of the need for
biological knowledge in medicine, agronomy and other
practical concerns. In addition, he may well feel that biology
helps us to understand ourselves, and that it is needed to
combat the myths which still surround many of our ideas
about birth, death, life, disease, health, and growth. These
attitudes or beliefs may well result in actions which reflect
his concern to use his knowledge for society—he may become
involved in the work of Natural History Societies, Field
Centres, or Conservation Corps, or in the education of parts
of the community about drugs or contraception. His
attitudes and actions form a consistent pattern of behaviour.

It may seem extravagant to suggest that the children to
whom we teach biology for a few short years should also
develop some of these characteristics. But education is
concerned with the future. We are educating *for* the
future—biological knowledge acquired in school is only of
value if it can be used *after* leaving school; inquiry skills

learned in school are only valuable if they can be used in *other* situations. Similarly, honesty, integrity, open-mindedness, independence and cooperation are qualities which we attempt to foster in children so that they continue to show these qualities as adults. We also hope that the curriculum can help them to develop consistency of belief and action.

Summary of arguments
In order to be able to justify the inclusion of biology in the curriculum, biology teachers must consider:

1 the unifying themes (or principles) of biology;
2 the inquiry skills which make biology a scientific discipline;
3 the values and attitudes of biology.

These considerations are a necessary preliminary to the statement (or selection) of aims (or goals) for a biology course. In turn, these aims are the basis for the selection of subject matter and teaching method. When these general aims have been chosen, it is necessary to transform them into more specific objectives (see chapter 8) before the actual lesson (or series of lessons) can be planned in detail.

The enunciation of aims and the production of objectives are both difficult and time-consuming, and there is a tendency to ignore them. Teaching without aims is like a ship without a rudder and, as Mager (1962) says, 'if you don't know where you're going, you're liable to finish up some-place else'. Thus it is unrealistic to suppose that any biology teacher will be able to clarify and express his aims at the first attempt. It *is* realistic to suggest that the development of a syllabus, subject-matter, resources and expertise will be a gradual process which should accompany—but not precede—the development of aims and objectives.

The selection of subject matter and teaching method will be influenced by many other factors, for example, the age and ability of the pupils, the resources and facilities of the school, and the pressures exerted by colleagues, parents and external examinations. These factors should not, however, affect the aims of the course. There are other factors which should influence both the translation of aims into objectives and the selection of content. These are:

the social implications of biology;
the relevance of biology to the everyday life of the pupil;
the extent to which biology makes a unique contribution
to the curriculum when compared with other disciplines.

Social implications and relevance
Many of the world's major problems—pollution and conservation, food production and starvation, overpopulation
and loss of natural resources—as well as many technical
advances—improved methods of birth control, transplantation of tissues and organs, the development of
machines which can replace natural organs—depend upon
biological knowledge. And yet these are precisely the issues
which find little or no place in the biology teaching of the
majority of secondary-school children.

In a similar way, problems which confront many
children—to smoke cigarettes or not, to experiment with
other drugs, to eat a balanced diet, to maintain a healthy
mind and body, to develop leisure activities—also have a basis
in biological knowledge. Again, many of these problems are
totally excluded from the planned curriculum although they
may play an important part in the informal education which
occurs outside the classroom. It may be that we need to
reassess the priorities which have been given in the past to
both biological facts and biological processes. The new
priorities might involve:

1 making the study of man and his relationships with
  other species the central concern of biology teaching in
  schools;
2 limiting our concern to 'useful' facts, ones which
  contribute to decisions about relationships between man
  and man, and man and his environment;
3 limiting our concern to 'useful' processes, ones which
  help individuals to understand the process of decision-
  making;
4 encouraging children to be involved in the making of
  both private and public decisions.

Two difficulties in adopting this approach are immediately

obvious: many of these personal and public problems require knowledge of other disciplines as well as biology; and many of the decisions involve ethical, or moral, or value-judgements. The latter area is one from which, traditionally, biologists— and other scientists—have tended to withdraw. Although it is often claimed that biology occupies a unique position between the natural and the social sciences (OECD, 1964), there is little evidence which suggests there is a growing appreciation of interconnections between these areas of knowledge. A biologist (shall we say) who steps out from his studies of behaviour in animals to speculate about the implications for human behaviour (psychology) may be savaged by both parties. Similarly, a psychologist who is a specialist in the measurement of human intelligence may be far from welcome if he reflects on the relationships between intelligence and heredity. The so-called 'popularisation' of science is also viewed with suspicion. In these cases, it may be that the 'community of scientists' (Cotgrove, 1967) exerts too strong a conforming influence on the activities of its members. If specialist scientists are reluctant to take decisions outside their subject areas, the school biology teacher may be even more reluctant. The belief that science is not concerned with value-judgements is an even bigger stumbling block for many biologists (but see chapters 12 and 13).

These difficulties raise two further questions: Is the separation of biology from other disciplines essential at the school level? Does biology make a unique contribution to the curriculum? The themes or principles of biology would appear to be capable of making a unique contribution, since the subject matter is concerned with living organisms and no other discipline has this as its main concern. The inquiry processes, however, are similar to those used in other scientific disciplines — chemistry, physics, geology, astronomy — and it may be that biology does not afford a unique training in these abilities, unless one concedes that the large number of variables involved in many biological investigations, and the consequent need for controlled experiments capable of statistical analysis, make such a contribution. The social sciences, psychology and sociology could make similar claims, however, with equal justification.

Finally, if we consider the kind of person the study of biology might produce—honest, integrated, open-minded, critical, independent, cooperative and evaluative—it might be expected that other disciplines, history and geography for example, would have products with similar attributes.

It is these kinds of questions, amongst many others, which have led to interdisciplinary experiments of various kinds.

Breakdown of disciplines

The General Science movement (SMA, 1950), plans for Health Education (Bibby, 1955) and the ideal of Education for Citizenship (Dewey, 1938; Mead, 1929) are early examples of such interdisciplinary endeavours. Their relative failure is probably due to a variety of factors, two of which were: they were unable to convince subject specialists that their aims were worthwhile; there was often uneasy juxtaposition of apparently unrelated subject matter. More recent attempts, Nuffield Combined Science (Nuffield, 1970), Social Studies (e.g. East Midlands CSE, 1973), and various Environmental Studies syllabuses (Carson, 1971; AEB, 1973), may have more chance of success; although the aims are still vaguely expressed, instructional objectives are lacking, and subject matter appears unrelated.

The rational case for interdisciplinary work is convincing, however, at many levels. At the subject-matter (or concept) level, does it make sense to teach a *biological* concept of energy, a *chemical* concept of energy and a *physical* concept of energy on different occasions? When studying the environment, are there distinctive chemical, physical, biological, geographical and sociological points of view which have to be taught separately?

At the inquiry level, is *chemical* observation somehow different from *physical* and *biological* observation? Is the ability to invent hypotheses a different activity in physical, biological and sociological investigations?

At the social problems level, 'answers' which ignore a multidisciplinary approach can only be partial ones. Drug addiction, for example, may be partly understood from knowledge of human physiology, but psychological and sociological factors, geographical and climatic conditions, and cultural norms are equally important in providing the 'useful' facts on which decisions are based.

The multi-disciplinary approach to research is a common feature of both pure and applied science, in universities, government departments and industry. It involves the co-operation of a variety of specialists, and hybrid scientists—biochemists and biophysicists—are often present in these teams. Perhaps the multi-disciplinary approach is now appropriate to learning and teaching. The main features of this approach might be:

1 it must be 'problem-centred' or 'issue-centred';
2 answers to problems (or issues) must be multi-dimensional;
3 answers must involve several appropriate disciplines;
4 students must be involved in the definition of the problem, as well as the production of answers.

Many teachers, of course, are already involved in this kind of team-teaching, although the extent to which it is (a) problem centred, and (b) involves self-initiated and self-directed activity by children is probably limited. Whether a teacher favours a subject approach or a multi-disciplinary one, the clarification of aims in terms of the contribution a discipline can make is essential, not only for selection of course content but also for evaluation of the work—although all three aspects of teaching (or curriculum development) will develop, overlap and interact with each other. Whichever approach a teacher favours, he will be working to a large extent in the dark. We simply do not know how these different approaches affect the children. We do not know how the new science projects, for example, affect the children (see chapters 14 and 16). Research is needed before conclusions can be made about the results of these curriculum developments.

This chapter has been mainly concerned with emphasising the need for biology teachers to clarify their aims either for teaching their own subject or when involved in multi-disciplinary work. Many other, equally important, problems have been ignored, for example, the relationship between science teaching and learning theory, the extent to which children are motivated by being involved in the planning of their own work, the need to change the ways in which teachers are trained.

Perhaps some uniformity of approach is needed which will encourage curriculum development, satisfy many of the psychological needs of children, and provide a basis for comparative studies.

The subversive approach

The required technique may already have been provided by Postman and Weingartner (1969). Certainly it could be used in a variety of situations. A subject teacher, like a biologist, could use it with classes of pupils at all ages and levels. A team of teachers could use it in multidisciplinary studies. It might also be used as a basis for in-service courses or conferences for teachers, or in the initial training of teachers. If the technique was used extensively, there would be a need for an organisation which could disseminate information about its use and provide help with problems of evaluation. If biology teachers were to use this technique, the formation of a Biology Teachers Association, either distinct from or part of the Institute of Biology, might be one way of providing such an organisation.

The instructions given by the teacher (or organiser, or leader) would be as follows. The blank space would be completed according to the subject of the course (which might be biology or environmental science or training for teaching).

Suppose all the syllabuses and curricula and textbooks in . . . have disappeared. Suppose all tests and examinations were lost. Then suppose that we have decided to turn this 'catastrophe' into an opportunity to increase the relevance of this . . . course to your needs and interests. What could we do?

We have a possibility for you to consider: suppose that we decide to have the entire course consist of questions. These questions would have to be worth seeking answers to from your point of view, and it might be particularly useful to choose questions which would help you in this rapidly changing world.

So let us try to compose a list of questions: questions that you are interested in and would like to find some answers to during the course. Make a list of twenty questions (more if you like) and then we will start.

This approach appears to be a radical one but it seems worth trying because:

1 it involves the students in the planning of their work, in the making of decisions, and shows your interest in their ideas;
2 the questions are likely to be 'problem-centred' or 'issue-centred' as well as 'information-centred', and show concern for social problems;
3 it is designed to increase the relevance of the course of work;
4 many of the questions will coincide with work that you had planned to include.

References

AEB., *Syllabuses 1973*, Environmental Studies, Associated Examining Board, Aldershot, Hampshire.

F. C. Bartlett (1951) *The Mind at Work and Play*, Allen & Unwin.

W. I. B. Beveridge, (1961) *The Art of Scientific Investigation*, Mercury Paperbacks.

C. Bibby (1955) *Health Education*, Heinemann.

S. McB. Carson (compiler), (1971) *Environmental Studies: The Construction of an 'A' level Syllabus*, National Foundation for Educational Research.

R. D. Cohen (1972) 'Problems of "retreading" Science Teachers' Part II, *Science Education*, 56 3.

S. F. Cotgrove (1967) *The Science of Society*, Allen & Unwin.

S. F. Cotgrove and S. Box (1970) *Science, Industry and Society*, Allen & Unwin.

J. Dewey (1938) *Experience and Education*, Kappa Delta Phi publications, (reprinted 1963, Collier-Macmillan).

East Midlands Regional Examinations Board, *Regulations and Syllabuses for the CSE, 1973*, Social Studies, Aspley, Nottingham.

J. E. Eggleston (1971) 'Biology', in *Disciplines of the Curriculum*, edited by R. C. Whitfield, McGraw-Hill.

L. Hudson (1970) *Frames of Mind*, Penguin Books.

F. R. Jevons (1961) *The Teaching of Science*, Allen & Unwin.

M. L. Johnson-Abercrombie (1960) *The Anatomy of Judgement*, Hutchinson.

J. F. Kerr (ed) (1968) *Changing the Curriculum*, London University Press.

R. F. Mager (1962) *Preparing Instructional Objectives*, Fearon.

A. R. Mead (1929) 'Qualities of merit in good and poor teachers', Journal of Educational Research, 20, 4.

P. B. Medawar (1964) 'Is the Scientific Paper a Fraud?' in *Experiment*, edited by D. Edge, BBC Publications.

National Society for the Study of Education (1932) *Thirty-first Year Book, Part I: Programme for Teaching Science*, University of Chicago Press.

Newsom Report (1963) *Half Our Future: A Report of the Central Advisory Council for Education (England)*, HMSO.

J. D. Novak (1965) 'A Model for the Interpretation and Analysis of Concept Formation', *Journal of Research in Science Teaching*, 3, 72.

Nuffield Combined Science, (1970) Longman/Penguin Books.

OECD (1964) *New Thinking in School Biology*, edited by L. C. Comber, OECD.

S. Postlethwait, J. Novak and H. Murray. (1969) *The Audio-Tutorial Approach to Learning*, Burgess.

N. Postman and C. Weingartner (1969) *Teaching as a Subversive Activity*, Dell, and in *Education for Democracy*, 1970, D. Rubinstein and C. Stoneman, (eds), Penguin Books.

C. E. Preston (1936) *The High School Science Teacher and his work*, McGraw-Hill.

A. Roe (1953) A Psychological Study of Eminent Psychologists and Anthropologists, and a comparison with Biological and Physical Scientists, Psychological Monographs, 67.

Science Masters' Association (1938) *The Teaching of General Science, Part II (Final Report)*, John Murray.

Science Masters' Association (1947) *The Teaching of Science in Secondary Schools*, (1st edition), John Murray.

Science Masters' Association (1950) *The Teaching of General Science* (1st edition), John Murray.

Science Masters' Association (1953) *Secondary Modern Science Teaching* (1st edition), John Murray.

C. S. G. Selmes B. G. Ashton, H. M. Meredith and A. B. Newall (1969) 'Attitudes to Science and Scientists', *School Science Review*, 51, 174.

C. P. Snow (1958) *The Search*, Macmillan.

P. H. Taylor (1970) *How Teachers plan their courses*, National Foundation for Educational Research.

B. E. Thompson and M. O. Pella (1972) 'A List of Currently Credible Biology Concepts Judged by a National Panel to be Important for Inclusion in K-12 Curricula', *Science Education*, 56, 2.

J. D. Watson (1970) *The Double Helix*, Penguin Books.

R. C. Whitfield (ed) (1971) *Disciplines of the Curriculum*, McGraw Hill.

S. Wiseman and D. Pidgeon (1972) *Curriculum Evaluation*, National Foundation for Educational Research.

Further Reading

S. A. Barnett (1971) *The Human Species*, Macgibbon & Kee.

C. Bibby (1959) *Race, Prejudice and Education*, Heinemann.

Conservation Trust (1973) *Education for Our Future*, The Conservation Society, Walton-on-Thames.

P. R. Erlich (1971) *The Population Bomb*, Pan Books.

E. Goldsmith (ed) (1972) *Can Britain Survive?*, Sphere Books.

G. Rattray Taylor (1969) *The Biological Time Bomb*, Panther Science.

G. Rattray Taylor (1970) *The Doomsday Book: Can the World Survive?* Thames & Hudson.

G. Rattray Taylor (1972) *Rethink: a Paraprimitive Solution*, Secker & Warburg.

A. Toffler (1971) *Future Shock*, Pan Books.

S. Wiseman and D. Pidgeon (1972) *Curriculum Evaluation*, National Foundation for Educational Research.

The Ecologist, Ecosystems Ltd., 11 Mansfield Street, Portland Place, London, W1M 0AH.

Your Environment, 10 Roderick Road, London, N.W.3.

Notes on Contributors

Michael Balls lectures on cell and development biology in the School of Biological Sciences, University of East Anglia. He has held research posts in the universities of Oxford, Geneva and California (Berkeley) and at Reed College, Oregon. He is the author of a number of articles.

Philip Booth is Head of Science and Mathematics at Hope High School, Flintshire, and was formerly Head of Biology at Thomas Bennett School, Sussex. He was joint organiser with Donald Reid of the project on Independent Learning in Biology set up by the Nuffield Foundation's Resources for Learning Project and is the co-author with him of a number of books and articles.

Marian Dawkins is a Research Officer with the Animal Behaviour Research Group, Department of Zoology, Oxford. She is the author or co-author of papers on the perception of food by birds and the control of behaviour sequences.

Wilfrid Dowdeswell is Professor of Education in the University of Bath. He was formerly the Senior Biology Master and Head of the Science Department at Winchester College. He is the author of four books and a number of research papers and articles, and organiser of the Inter-University Biology Teaching Project, the Nuffield O-Level Biology Project and also (with Dr P.J. Kelly) of the Nuffield A-Level Biology Project.

Brian Dudley is Lecturer in Biology Education in the University of Keele. He is a member of the Joint Biological Education Committee of the Royal Society and Institute of Biology and has written chapters of the revised Nuffield O-Level Biology Project.

Jack Dunham is Lecturer in Psychology and Student Counsellor in the University of Bath. He has been, among other things, a PE teacher, a teacher of backward children and an educational psychologist. His research work and publications have included studies of childrens' and students' attitudes.

Doris Falk was, from 1956–71, Professor of Biology at

California State University at Fresno and is now writing and acting as a consultant for Fresno City and County schools. She was a member of the State of California Science Advisory Committee and Director of two National Science Foundation Institutes for high school biology teachers.

Dennis Fox is Principal Lecturer in Biology at Nottingham College of Education. He was formerly at Queen Elizabeth Grammer School, Wakefield, has written various articles and contributions to books and was a team member of the Nuffield Secondary Science Project (1966–70) and the Nuffield 16+ Science Project (1972–73).

Alexander Geddes is Lecturer in the Astbury Department of Biophysics in the University of Leeds. His research, which includes a year at Indiana University Medical Centre, has been largely in the field of x-ray crystallographic studies on fibrous proteins, drugs and hormones.

David Hornsey is Lecturer in Radiobiology in the University of Bath. He is the author of various articles and a book, and co-author of another.

Hugh Iltis is Professor of Botany and Director of the Herbarium at the University of Wisconsin in Madison. His special interests are in problems of plant distribution and evolution, the origin of maize, and man's optimum environment and need for nature. He has led botanical expeditions to Costa Rica, Mexico and Peru and has published widely in scientific and popular journals. Much of his time recently has been spent in lecturing on the urgent need for the preservation of biotic communities, and criticizing agricultural practices, especially the 'Green Revolution'.

Jane Jenks teaches biology at Lady Margaret School, London. She has recently finished writing the handbook and commentaries for a series of sex education films.

Richard Joske is Professor of Medicine in the University of Western Australia. He has published many papers dealing chiefly with diseases of the alimentary system and clinical immunology. He has studied in London and Massachusetts as well as Australia.

Robert Lister is Senior Lecturer at the University of London Institute of Education in the Department of Biological

Science and has taught at several schools. He is the author of various papers and was responsible for the trial and development of the assessment techniques used in the Advanced Level Nuffield Biology Examination.

Donald Reid is Assistant Director of Education and Training with the Health Education Council, London, and was formerly Head of Science at Thomas Bennett School, Sussex. He is co-author with Philip Booth of a number of books and articles and was joint organiser with him of the project on Independent Learning in Biology set up by the Nuffield Foundation.

Cyril Selmes is Lecturer in Education and Director of Studies for the Postgraduate Certificate Course in the School of Education, University of Bath. He has taught in various schools, secondary modern, grammar and grammar-technical, and during this time he was Head of a Science department for seven years. His first degree included both Biological Sciences and Experimental Psychology and for his PhD he studied attitudes towards science.

Colin Stoneman has spent fourteen years teaching biology in secondary schools, from which he was seconded to the Nuffield Advanced Biological Science Project. He has been six years as lecturer in the Department of Education, University of York. He has written material for the Project, a topic book for schoolchildren, and various articles.

David Tomley is Lecturer in the School of Education at Leicester University. He was previously Senior Lecturer in Biology at the City of Portsmouth College of Education and has also taught in schools. He was a trials course teacher of the Nuffield O and A-Level Biology Courses and a Joint Area Coordinator.

Michael Tribe is Lecturer in Biological Sciences at Sussex University where, in 1969, he was appointed Director of the Inter-University Biology Teaching Project. He has spent four years as a schoolteacher and has a wide interest in curriculum innovation in both secondary and tertiary education.

Peter Whittaker is Lecturer in Biochemistry and Sub-Dean of Biological Sciences at Sussex University. He was previously Lecturer in Genetics in the Botany Department at Hull University. He is the author and co-author of a number of papers, and a book with Michael Tribe.

Acknowledgements

In Chapter 1, the photographs are by courtesy of Dr C. Hackenbrock (1968), J. Cell. Biol. 37, 345; in Chapter 2, the electron micrograph is by kind permission of Professor R.D. Preston; in Chapter 5, figure 2 is by kind permission of Methuen Educational Ltd and Dr. E.B. Ford; in Chapter 6, figure 1 is by kind permission of Methuen Educational Ltd and W.M.M. Baron; in Chapter 9, figure 3 is from *Movement in Animals* (Book 3 of *Biology for the Individual* by Donald Reid and Philip Booth) and figure 4 is from *Some Problems of Life in Hot, Cold and Dry Climates* (Book 5 of the same series) by kind permission of Heinemann Educational Books.

In Chapter 13, quotations from Schools Council Working Paper No 1, *Science for the Young School Leaver*, (1965) are by kind permission of the Schools Council.

In Chapter 18, lists of principles of biology are by kind permission of McGraw-Hill and C.E. Preston; University of Chicago Press and the National Society for the Study of Education; John Wiley & Sons Inc., B.E. Thompson and M.O. Pella.

Dr A. Geddes wishes to acknowledge the valuable discussions with Professor Preston about the teaching of biophysics in secondary schools.

Dr M. Balls wishes to acknowledge the support of the Medical Research Council in his research on the control of cell proliferation, differentiation and function *in vitro*. He is also grateful to his colleagues, Dr M.A. Monnickendam, Dr R.H. Clothier and Mr R.T.S. Worley, for their comments on Chapter 5.

Professor Iltis wishes to thank George B. Van Schaack, Dan McKinley, Tim Clark, Marylin Bagley, Ralph Buchsbaum, Fred Iltis and Joyce Gutstein for valuable comments.

Dr C. Selmes wishes to acknowledge the helpful comments made by Professor S.A. Barnett on the attitudes of biologists. He is also grateful to his colleagues, Professor W.H. Dowdeswell and Dr G.H. Hones, for their comments on Chapter 18.